Total Quality Control Essentials

Other McGraw-Hill Books of Interest

BERRY • *Managing the Total Quality Transformation*
CARRUBA • *Product Assurance Principles Integrating Design and Quality Assurance*
CROSBY • *Let's Talk Quality*
CROSBY • *Quality Is Free*
CROSBY • *Quality without Tears*
FEIGENBAUM • *Total Quality Control Revised 40th anniversary edition*
GRANT AND LEAVENWORTH • *Statistical Quality Control*
HARRINGTON • *Business Process Improvement*
HRADESKY • *Productivity and Quality Improvement*
JURAN AND GRYNA • *Juran's Quality Control Handbook*
MILLS • *The Quality Audit*
OTT AND SCHILLING • *Process Quality Control*
ROSS • *Taguchi Techniques for Quality Engineering*
SLATER • *Integrated Process Management: A Quality Model*
TAYLOR • *Optimization and Variation Reduction in Quality*
TAYLOR • *Quality Control Systems*

Total Quality Control Essentials

Key Elements, Methodologies, and Managing for Success

Sarv Singh Soin

McGraw-Hill, Inc.
New York   St. Louis   San Francisco   Auckland   Bogotá
Caracas   Lisbon   London   Madrid   Mexico   Milan
Montreal   New Delhi   Paris   San Juan   São Paulo
Singapore   Sydney   Tokyo   Toronto

Library of Congress Cataloging-in-Publication Data
Soin, Sarv Singh [Sarvandan S.]
   Total quality control essentials : key elements, methodologies, and managing for success / Sarv Singh Soin [Sarvandan S.].
       p.   cm.
   Includes index.
   Includes bibliographical references.
   ISBN 0-07-059548-8
   1. Total quality management. 2. Quality control.   I. Title.
   HD62.15.S63   1992   92-7968
   658.5'62—dc20   CIP

Copyright © 1992 by McGraw-Hill, Inc. All rights reserved. Printed in the United States of America. Except as permitted under the United States Copyright Act of 1976, no part of this publication may be reproduced or distributed in any form or by any means, or stored in a data base or retrieval system, without the prior written permission of the publisher.

2 3 4 5 6 7 8 9 0   DOC/DOC   9 8 7 6 5 4 3

ISBN   0-07-059548-8

The sponsoring editor for this book was Gail F. Nalven, the editing supervisor was Nancy Young, and the production supervisor was Pamela A. Pelton. This book was set in Century Schoolbook by McGraw-Hill's Professional Book Group composition unit.

Printed and bound by R. R. Donnelley & Sons Company.

> Information contained in this work has been obtained by McGraw-Hill, Inc., from sources believed to be reliable. However, neither McGraw-Hill nor its authors guarantees the accuracy or completeness of any information published herein and neither McGraw-Hill nor its authors shall be responsible for any errors, omissions, or damages arising out of use of this information. This work is published with the understanding that McGraw-Hill and its authors are supplying information but are not attempting to render engineering or other professional services. If such services are required, the assistance of an appropriate professional should be sought.

Contents

Foreword xi
Preface xiii

Chapter 1. The Quality Revolution — 1

Overview—The Genesis of Total Quality Control — 1
Benefits of Quality — 2
  PIMS Study — 3
  Deming Prize Data — 3
  Specific Data from One Company — 4
  Benefits of a TQC Effort — 5
Objective of a Total Quality Control Effort — 5
  Definition of Quality — 6
Need for a System — 6
  Key Elements of TQC — 8
Questions and Answers on the Quality Revolution — 9
Summary: The Quality Revolution — 10
References — 11

Chapter 2. Customer Obsession — 13

Overview — 13
Customer Obsession via a Systematic Approach — 15
Customer Complaint and Feedback Management System — 15
  Need for a System to Manage Customer Complaints — 17
  Customer Complaint and Feedback System — 18
  Promoting and Facilitating the System — 22
  Trends in a Customer Complaint and Feedback System — 25
  Summary: Customer Complaint and Feedback System — 27
Customer Satisfaction Surveys — 27
Conducting Customer Satisfaction Surveys — 28
  Setting Priorities for Issues and Customer Comments — 29
  Customer Satisfaction Model — 30
  Other Types of Customer Satisfaction Data — 32
  Ensuring Follow-Up to Customer Satisfaction Surveys — 33

vi   Contents

Competitive Benchmarking	33
Purpose of Benchmarking	33
What Can Be Benchmarked?	34
Benchmarking Process	34
Capturing the Customer's Voice and Needs	38
Why Do Products Succeed in the Marketplace?	39
Quality Function Deployment	42
Meeting and Exceeding Customer Needs	43
Attractive Quality in Products	43
Attractive Quality in Services	43
Questions and Answers on Customer Obsession	45
Summary: Creating a Customer-Obsessed Organization	46
References	46

Chapter 3. The Planning Process   49

Overview	49
Essentials of a Planning Process	50
Long-Range Plan	51
Annual Plan	53
Hoshin Planning Process	54
Daily Management, or Business Fundamentals, Plan	56
Launching the Plan	56
Hoshin Plan Reviews	56
Hoshin Plan: Ensuring Success	56
Relation between Hoshin Plan and Daily Management Plan	58
MBO or Hoshin Planning?	58
Formats and Guidelines	58
Annual Hoshin Plan	58
Chief Executive's Objectives	64
Deployment and Cascading of Objectives	68
Implementation Plan	72
Daily Management, or Business Fundamentals, Plan	74
Control Limits	78
Setting Numerical Targets	79
Review for Hoshin and Daily Management, or Business Fundamental, Plans	80
Planning: Putting It All Together	85
Planning Calendar	85
Planning for Special or R&D Projects	85
Planning and Budgeting	87
Questions and Answers on Hoshin and Daily Management, or Business Fundamentals, Planning	87
Summary: The Planning Process	90
References	91

Chapter 4. The Improvement Cycle   93

Overview	93
Selecting Items to Improve	94
Cost of Poor Quality	95
PDCA Cycle	96
The Shewhart and Deming Cycle	97
The PDCA Cycle	97
The CA-PDCA Cycle	97

QC Story	97
Modified and Improved PDCA Cycle	98
Relationship between Improvement and Control	98
Benefits of the PDCA Improvement Cycle	99
Detailed PDCA Cycle	100
Plan Stage	100
Do Stage	106
Check Stage	108
Act Stage	109
Seven Quality Control Tools and Other Methodologies	111
Education of Employees	112
Source of Errors and Defects	114
Human Errors	114
Where Can You Use the PDCA Cycle?	114
Example in PDCA Format	116
Standards	118
Benefits of Standards	120
Important Points for Creating Standards	120
Preventing Recurrence	120
Problem-Solving Hierarchy	122
Questions and Answers on the Improvement Cycle	124
Summary: The Improvement Cycle	125
References	126

Chapter 5. Daily Process Management   127

Overview	127
Daily Process Management	128
The Process Concept	128
The Requirements for Daily Process Management	129
Prerequisites for Good Process Management	130
Identifying Key Processes to be Managed	132
Competitive Benchmarking	133
Product Development Process	133
Introduction to QFD	136
Review of QFD Basics via the Pencil Example	141
Developing the Customer's Voice for QFD	147
Recommendations for Increasing QFD Success in Your Company	149
Summary of QFD	154
Design Standards	154
Thermal Design and Measurement	155
Component Derating	155
Specific Electrical and Mechanical Standards	156
Supplier Management	156
Failure Mode Effects Analysis	157
Analysis of the New Product Development Process via the Project Postmortem and T-Type Matrix	160
The Project Postmortem	161
The T-Type Matrix	167
Summary of Project Postmortem and T-Type Matrix	178
Quality Assurance System	178
Quality Assurance Improvement Cycle	181
Quality Checkpoints Chart and System	181
The Sales Process	187
The Sales Funnel	187

The Sales Situation	190
Identifying Appropriate Performance Measures	190
Sales Won or Lost Postmortem	193
Summary of Sales Process	198
Managing Other Processes	199
Reducing Design Time or Time to Market	200
Questions and Answers on Daily Process Management	201
Summary: Daily Management Process	202
References	203

Chapter 6. Employee Participation — 205

Overview	205
Quality Circle, or Team, Activity	206
Management of Quality Circle, or Team, Activity	206
An Important Afterword on Quality Circles and Teams	211
Employee Suggestion Schemes	211
Guidelines for an Employee Suggestion Scheme	211
Promotion and Measurement of Activity	216
Education in Quality Methodologies	219
Publicity, Promotion, and Recognition	220
Some Questions and Answers on Employee Participation	221
Summary: Employee Participation	226

Chapter 7. Getting Started and Ongoing Management — 227

Overview: Why Start a TQC Effort?	227
The Quality Steering Committee	228
Establishing a Quality Vision and Goal	229
Getting Up a TQC Headquarters Function	230
Phases of a TQC Effort	230
Introduction Phase	231
Acceleration Phase	231
Cruising Phase	232
Second Accelerationn Phase	232
Second Cruising Phase	233
TQC and Time Management	233
The Basic Activities of a Manager	233
Fitting the TQC Elements in the Itoh Model	235
Some Questions and Answers on Getting Started and Some General Questions on TQC	237
Summary: Getting Started and Ongoing Management	239

Chapter 8. Conducting TQC Audits, or Reviews — 241

Overview	241
Deming Prize and Malcolm Baldrige Award Criteria	242
Deming Prize Criteria	242
Malcolm Baldrige National Quality Award	245
Comparison between the Deming Prize and the Malcolm Baldrige Award	246
A TQC Review Procedure	247
The TQC Review Process	247
An Agenda for the Review	248
Detailed Checklist for the Review	250

Scoring System	250
Review Checklist	252
The Review Team	267
Preparing and Issuing the TQC Review Recommendations	269
Common Problems Discovered during the Review Process and Some Review Guidelines	269
Presidential Audits, or Executive Reviews	274
Review Agenda for Presidential Audit	275
Issuing Recommendations after a Presidential Audit, or Executive Review	278
Summary: Conducting TQC Reviews	278

Chapter 9. Some Thoughts on the Essence of TQC — 281

Quality First	281
Continuous Improvement	281
Attractive Quality	282
Supporting Methodologies and Techniques	282
Customers First	282
Supporting Methodologies and Techniques	283
The Importance of the Way or Process	283
Balance between Standards and Creativity	283
Supporting Methodologies and Techniques	283
The Organization That Learns and Grows	283
Supporting Methodologies and Techniques	284
TQC: Organization and People Running at Peak Efficiency	284
TQC, Creativity, and Success	284

Conclusion — 285

Appendix 1. Improvement Project to Reduce Dissatisfied Customers Using the PDCA Cycle — 287

Appendix 2. The Seven Quality Control Tools and the New Seven Tools — 297

Bibliography — 305

Index   307

Foreword

Today, in the midst of this country's longest and deepest recession in the last four decades, it is perhaps more obvious than ever that a business, if it is to survive or prosper, simply must embrace the philosophy and methodology of total quality control (TQC).

But even those who recognize the need struggle to understand and implement the changes in personal behavior and company culture that are required. This condition exists because, in spite of the fact that many books and articles have been written on the subject, few truly practical guides exist. This book represents such a guide and will be an invaluable tool to those dedicated to the required transformation or change of state.

The book is Soin's second. The first, *Total Quality Control at Hewlett-Packard—The Asian Experience*, was published in 1986 and was distributed widely within the company. The book facilitated the transfer of the methodologies first adopted and used so successfully by Hewlett-Packard's Asian operations to the rest of the company (Yokogawa Hewlett-Packard, the company's subsidiary in Japan, won the Deming Prize in 1982).

This book is an extension of the work. In it, Soin details the management system used by Hewlett-Packard to measure and accelerate its progress toward true and sustained implementation of TQC. The first few chapters provide details for each of the five basic elements of TQC that comprise the complete system.

The eighth chapter provides a description of a review or audit methodology. Called a "TQC Review" within Hewlett-Packard and first used in 1988, the methodology is central to the company's current program. The reviews have served to demonstrate the relevance and importance of TQC to general management in a very practical way.

The review method has done more to accelerate the company's progress toward TQC than any other approach adopted during the last decade.

This book is an unusual combination of theory and practical example. It provides a translation of philosophy to real application that's both comprehensive and easily understandable. It is an invaluable addition to the existing literature on total quality control.

Craig Walter
Corporate Quality Director
Hewlett-Packard
January 18, 1992

Preface

In the past few years there has been a surge of books on quality, new quality methodologies, and new acronyms such as TQC, TQM, QFD, ad infinitum. This blizzard of ideas and methods is an indication of an increased interest in quality. In part this is due to the Japanese wave of high quality products and services that are attractive, satisfying, and affordable. In fact, the secret of Japanese success in world markets can be found "on the shelves and showroom floors all around the world: good quality products that people want and in such variety that any consumer whim can be satisfied," according to Akio Morita of Sony.
Many of the ideas that result in such good quality and satisfying products are captured in, or are part of, what is now called total quality control (TQC).

While TQC may be less important for a young and nimble start-up company, which relies on one innovative product or technology, it is vital for a larger company or organization that needs to better focus its efforts and increase its effectiveness and efficiency.

This is a book that will help you implement the essentials of total quality control in your organization. It will help a design, manufacturing, sales, or service organization to launch a TQC effort. If you have already begun a TQC effort, it will introduce techniques that will help you go beyond what you have in place. In addition, we introduce a TQC review or audit procedure that will allow you to diagnose your current effort and improve it.

This book will also introduce or restate current and new quality initiatives and will organize them into a TQC system—a system that will run with management guidance and ensure the continuous generation of the highest quality products and services. An overall TQC system is important because new quality initiatives should not be introduced in an ad hoc manner; otherwise there will be danger of confusion or overload. Equally important, a good system will provide a strong foundation for the competitive and turbulent challenges of the 1990s.

How to Use This Book

Chapter 1 starts with an introduction to quality and lists the elements of TQC. Chaps. 2 through 6 give details on each of those elements. Chap. 7 gives advice on starting a TQC effort, while Chap. 8 provides a tested and effective method of auditing or reviewing a TQC effort. Chap. 9 gets into the essence and guiding principles of TQC.

All the chapters, except Chap. 5, are relevant to knowledge workers, professionals, managers, and chief executives involved in design and manufacturing, marketing and sales, service, multinational operations, management, government, or any other organization. Chap. 5 is very specific; it provides information on identification and management of key processes in a design, manufacturing, and sales company. If you work in a different environment, you will need to apply the concepts put forward in Chap. 5 to manage key processes in your environment. Chap. 8 provides a method for auditing or reviewing a TQC effort. This chapter will be used extensively for conducting formal and informal TQC reviews. An experienced quality manager should guide the effort in this area. By experienced we mean with many years working as a quality manager, with exposure to quality efforts at other companies, statistical quality control (because many TQC concepts originate from there), and an international environment. In the longer term, we recommend that chief executive officers or company presidents conduct the reviews—specifically executive reviews or presidential audits, which are discussed in Chap. 8.

The overall structure of the book can, and should, be used to formulate a total quality control education program. There is sufficient material here for the beginner, the professional, the consultant, all managers, and the chief executive. This material will have to be supplemented with the seven quality control tools and the seven new quality control tools and, for a design and manufacturing environment, with other statistical tools.

A final point. The purpose of this book is not meant to give you a quick fix in order to help you win a quality award. Instead it will ensure that you have a robust quality effort and high efficiency, resulting in increased business success and higher customer satisfaction. Having got that, winning a national quality award such as the Malcolm Baldrige award or Deming prize will be easy.

Onward.

Sarv Singh Soin

Acknowledgments

I am extremely indebted to the numerous people who have contributed to, or guided me in, my effort to write this book. Katsu Yoshimoto for his inputs over the years, Tan Bian Ee and Paul Ow for permission to use material from Hewlett-Packard; special thanks to Khushroo Shaikh and Mike Ward for editing. Craig Walter for his encouragement, Neoh Kah Thong for his valuable inputs, Walt Sousa for his continual inspiration, Dr. Noriaki Kano for his invaluable insights and permission to use some of his material, and Mona for her infinite patience and encouragement.

Total Quality Control Essentials

Chapter 1

The Quality Revolution

The right quality and uniformity are foundations of commerce, prosperity and peace.
W. EDWARD DEMING

Overview—The Genesis of Total Quality Control

Quality! Quality is on everyone's lips these days because it can make the difference between success and failure in a very competitive and tumultuous world. Today quality means more than product reliability; today it means a Total Quality Control (or TQC) effort—an effort in which everybody and every function in an organization participates.

The term *TQC* originated in the United States, but the current concept of TQC developed in Japan. After the Second World War, Japanese managers realized that they had to export or perish. And export they did, but the quality of their products was shoddy. Two renowned American statisticians and quality experts, Edward Deming and Joseph Juran, spent several years educating the Japanese on improving quality. Japanese product quality improved; the Japanese then went on to develop the current concept of TQC, which goes beyond the ideas of these great teachers to embrace quality in the entire Japanese company—in manufacturing, administration, marketing, sales, after-sales support, business planning, and services.

The introduction of the prestigious Deming Prize in Japan in 1951, awarded to Japanese companies that show a high commitment to quality, helped to motivate many Japanese manufacturing and service companies to improve their quality. Since then many other countries have introduced similar quality awards to encourage quality awareness and improvement in their industries.

Keen competition inside and outside Japan has allowed many companies throughout the world to improve. Today Japan is renowned for its manufacturing skills and quality. Failure to learn from Japanese companies and from other successful companies will greatly reduce the competitive ability of any company.

Let us start our discussion by reviewing the benefits of quality.

Benefits of Quality

What are the benefits of TQC? Many people will speak enthusiastically about the benefits of quality. But what is the value of high quality to the business enterprise? Let us look at a generic model of TQC and then review some industry data.

TQC that is properly implemented will focus on improving products, services, and processes; when these improve, they will have an impact on productivity, customer satisfaction, and profits. The impact will be seen both internally and externally, as illustrated in Fig. 1.1, and in the following:

Internally. When quality improves, we will get higher productivity, which allows us to lower prices (in which case we are competing with price), market share increases, and we get higher profits. Alternatively, the lower cost gives a direct increase in profits.

Externally. Higher quality allows us to increase customer satisfaction, increase customer loyalty, and get more repeat purchases. This results in increased market share and higher profits. Alternatively, we could compete on a value basis, charging a higher relative price for our higher quality. Many companies do this, for example, Sony in Japan, Boeing Aircraft in the United States, and Mercedes-Benz in

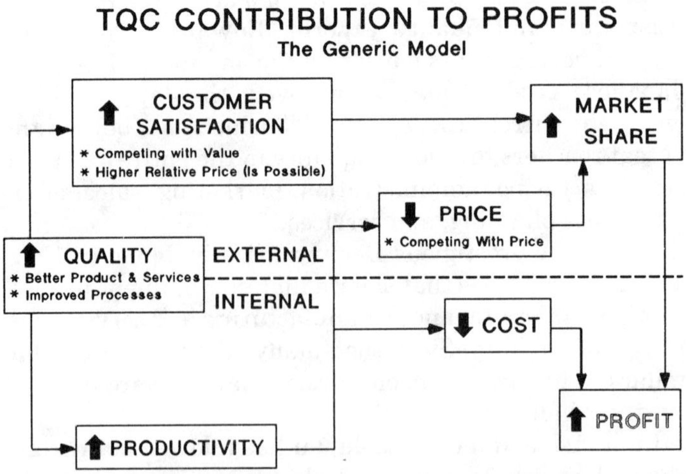

Figure 1.1 How TQC contributes to profits.

Germany. The result, again, is higher profits, although market share may remain about constant.

PIMS study

Let us look at some data on the impact of quality. Figure 1.2 shows the Profit Impact on Marketing Strategy (PIMS) study done by Buzzel and Gale.[1] This data from the Strategic Planning Institute shows the relationship between market share, relative quality, and return on investment (ROI). You can see that increased market share gives a higher ROI; you probably expected that. But an increase in relative quality also gives a higher ROI. The best situation is when both high market share and high quality exist—resulting in the highest ROI. This study supports our generic model.

Deming prize data

Let us look at another study. We earlier mentioned the Deming prize in Japan. A comparison of profits generated at Deming Prize and non-Deming Prize manufacturing companies is shown in Fig. 1.3.[2] It shows the operating profit of companies that have won the Deming Prize versus all other manufacturing companies in Japan. The data is cumula-

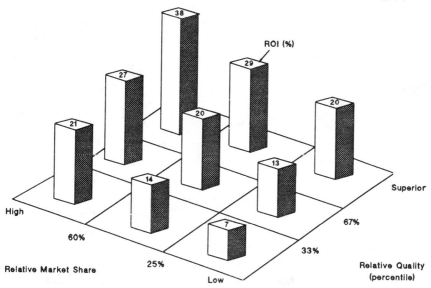

Figure 1.2 The impact of quality and market share on profits. Reprinted with permission of Free Press, a division of Macmillan, Inc. from *The PIMS Principles: Linking Strategy to Performance* by Robert D. Buzzel and Bradley T. Gale. Copyright © 1987 by The Free Press.

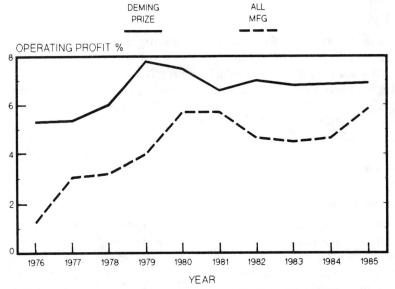

Figure 1.3 Profitability of Deming Prize companies. Data from JSQC paper, Dr. Kano, 1987.

tive—as more companies join the Deming Prize list, they are added in. You can see the spread; the Deming Prize companies are doing better.

The profit gap between the two sectors appears to be narrowing. In part, this is because many Deming Prize companies impose strict quality standards on their suppliers, who may have not won the prize. In addition, the increased global competition faced by non-Deming Prize companies is forcing them to improve quality. Finally, there is the impact of sharing experiences and learning between Japanese companies.

Specific data from one company

Let us now look at one company's quest for quality. In 1982, Yokogawa Hewlett-Packard (YHP), the Japanese subsidiary of the Hewlett-Packard Company, won the coveted Deming Prize. John Young, the CEO of Hewlett-Packard, quotes the following gains for YHP during a 5-year period (as reported in the *Journal for Business Strategy* and Hewlett-Packard Company presentations): "This is an excellent example of how a strategic focus on total quality can maximize the attainment of a company's key objectives—profit and growth." These improvements are shown in Fig. 1.4.

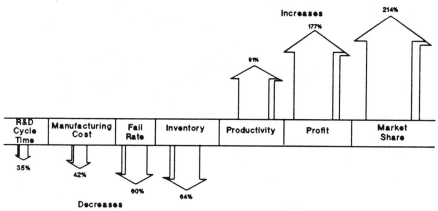

Figure 1.4 Improvements at YHP.

Benefits of a TQC effort

By now the data should be convincing. The question you may have is, How do I make this work for my company? That is the purpose of this text and we will go into great detail, but first let us list some of the benefits of a good TQC effort:

- Higher employee morale
- More efficient processes
- Higher productivity
- Less fire fighting, resulting in more time for innovation and creativity
- Improved quality of products and services
- Increased market share
- Lower costs
- Increased customer satisfaction
- Higher profits

Given some of the data we just reviewed, quality has to be part and parcel of a company's business strategy. This is the key to survival and growth in the 1990s— there is no alternative.

Objective of a Total Quality Control Effort

There are many definitions of TQC, but here's one definition that includes the objective of TQC:

An effort of continuous quality improvement of all processes, products, and services, through universal participation, that results in increasing customer satisfaction and loyalty, and improved business results.

In our context TQC goes beyond traditional product quality and also includes efficiency, productivity, customer satisfaction, and good management of key areas, such as planning and human resources.

Definition of quality

At this juncture it would be appropriate to review some definitions of quality:

- A definition that gained much popularity in the 1980s was Phil Crosby's "conforming to specifications." The difficulty with this definition is that the specifications may not be what the customer wants or is willing to accept.
- A much better definition is one that Joseph Juran has proposed for many years: "fitness for use." Fitness is to be defined by the customer.
- Noriaki Kano and others have proposed a two-dimensional model of quality.[3] Refer to the discussion in the box on page 7. It suggests that quality has two dimensions which are "must be quality," or a set of expected features, such as reliability, and, "attractive quality," or the unexpected that goes beyond the customer's needs; these are extra features that the customer would love to have but has not desired because he or she has not yet thought about them.
- By now we can come up with a comprehensive, yet simple, definition of quality such as "Products and services that meet or exceed customers' expectations."

Need for a System

Recently there has been a blizzard of new quality initiatives and methodologies. Introducing them ad hoc without a basic TQC system in place would result in a less than optimum benefit. Even without new initiatives, all ongoing quality efforts—product quality, improvements, processes, employee morale, customer issues and needs, planning, and a host of other issues—must run like clockwork. This is even more crucial in today's turbulent times and highly competitive environment. Small start-up companies with unique or innovative products do not

The Two Dimensions of Quality

Noriaki Kano and others[3] have proposed the concept of two dimensions of quality: "Must be quality" and "attractive quality." See Fig. 1.5.

"Must Be Quality" is that aspect of a product or service which the customer expects. If the customer does not get it, there will be extreme dissatisfaction. Examples of this are a reliable, safe, and easy to use product. This is the minimum acceptable standard—very like Joseph Juran's "fitness for use" concept.

"Attractive Quality" is that aspect of a product or service that goes beyond current needs. If a special feature is available, the customer will be thrilled and excited, but if this feature is not available, the customer has no comment. An example of this could be a sun roof, antilock brakes, or a safety air bag in a car. With time, such "attractive quality" becomes "must be quality." An example is the safety air bag which is fast becoming a "must be quality" item. Other items that were once "attractive quality" are remote control and multiprogramming. They are now standard features in most video color recorders (VCR's). So, higher quality is a never ending quest.

A well-designed product or service should have both dimensions of quality—these can strongly influence the customer's buying decision.

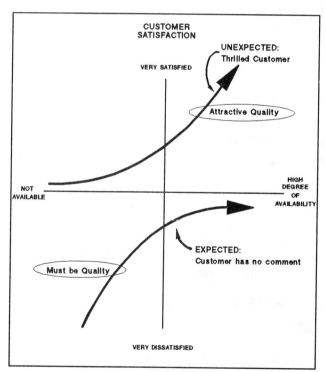

Figure 1.5 The two dimensions of quality.

need a system, but big organizations do. The alternative is inefficiency, complacency, and declining performance—just look at the current demise and incredible waste at many large corporations.

In the following pages we will elaborate on what is a good TQC system and how to manage it. But let us first determine the elements of TQC.9

Key elements of TQC

There are many ways to cut the TQC pie. Let us look at the recommendations of some experts. A. V. Feigenbaum in his book *Total Quality Control*,[4] first printed in 1951, suggests a system that manages the following:

1. Product quality control
2. Incoming material control
3. Manufacturing process control
4. Quality costs

This specific approach tends to limit TQC to the shop floor and trained specialists, but nevertheless it is an excellent way to ensure product quality.

Dr. Ishikawa, the father of Japanese TQC, once listed these as the basics of TQC:[5]

1. The next process is your customer.
2. Use of facts and statistics.
3. The plan do check act (PDCA) cycle.
4. Everybody participates.
5. Management commitment.

Dr. Ishikawa goes on to describe TQC as "thought revolution in management."[6] Other experts describe TQC as scientific management or management by facts.

We need an approach that is simple and yet emphasizes the items mentioned above. Additionally we must ensure that we have product quality, the right product, a competitive product, and a strong organization. Here is our approach to separating the various elements of TQC:

1. *Customer obsession.* This will include all activities required to keep your customers happy, satisfied, and—whenever possible—thrilled.

2. *The planning process.* This is the best way to show and implement management commitment to customers, employees, improving quality, and planning for the future. This is surely one of the most important processes in any organization.
3. *The improvement cycle.* This is to ensure a rigorous, effective, and systematic method of improving processes or reducing problems. We will propose the use of the PDCA improvement cycle.
4. *Daily process management.* This will ensure good day to day management of your key processes, resulting in efficient and predictable processes. The result will be a lower cost and a more efficiently managed organization.
5. *Employee participation.* All employees must be educated in quality techniques, ensuring a high degree of participation. Additionally, management must direct everybody and the organization toward a common goal.

The separation between elements is not clean, nor can it be, but it is sufficient for our discussion. In particular, the planning process, the improvement cycle, and daily process management are linked closely, while all other items will overlap. In the remainder of this text, we will discuss these five elements and will go a little beyond them.

Questions and Answers on the Quality Revolution

Q1. I can understand starting or increasing the total quality effort at a company or an organization that is doing poorly, but why should a successful or profitable company start a TQC effort?

A1. It must—to prevent complacency and to increase competitiveness. Today's successful company may be tomorrow's poor performer. Examples abound, let me mention a few:

Look at IBM, they were extremely successful in the 1970s and 1980s. Last year (1991) IBM had its first annual loss—of $2.8 billion. John Akers, COE of IBM, admits that IBM has lost touch with its customers.

Or look at Xerox, an extremely innovative and successful company. Xerox developed photocopier products and held a large market share in big machines. But it became complacent and allowed nimble Japanese competitors to enter the neglected small photocopier machine business. In the last decade these Japanese competitors have threatened Xerox.

The automobile industry is another example. The North American automobile manufacturers were very successful, but look at what is happening to them today.

Q2. Where did some of them go wrong?

A2. What are the long-term objectives of a large corporation? You will say: to make profits, increase market share, and so on. One objective, that I think is important, is to survive and remain viable long term. This may not be written down, but it is there.

But how many successful corporations plan for the long term? Japanese manufacturers planned to enter the North American and European market in the 1950s and 1960s. As their market share grew, what was the response? Specifically, what did General Motors (GM) do? Even as GM's market share declined—especially in the 1980s—it squandered its opportunities. It was cash rich, and it invested $90 billion in dubious acquisitions and investments instead of increasing its competitiveness. Do you know that GM tries to sell left-hand drive cars in Japan, where the norm is right-hand drive cars? And then it complains about market barriers when sales are miserable. Where was GM's vision? Where was its understanding of customer needs or the competition? What was its plan to respond to the never ending loss of market share? GM was fat and complacent. But even successful Japanese companies are not immune. Honda, the automobile manufacturer that is extremely successful in North America, does poorly in the Japanese market. Clearly, they will need to respond or lose that market. TQC can help successful companies stay successful.

Q3. How can TQC help successful companies?

A3. TQC methodology will help a company better understand its competition and keep it close to the customer. It also helps a company to plan better—both for the short term and the long term. It ensures a focus on priorities. Most important, TQC reviews—which we discuss in great detail—will help in providing regular diagnosis of a company and highlight its strengths and weaknesses, and it will provide recommendations for improvement.

Summary: The Quality Revolution

We have discussed the benefits of a TQC effort. A short list of the benefits includes the following:

- Higher employee morale
- More efficient processes
- Higher productivity
- Less fire fighting, resulting in more time for innovation and creativity
- Improved quality of products and services

- Increased market share
- Lower costs
- Increased customer satisfaction
- Higher profits

We defined quality as "Products and services that meet or exceed customers' expectations." We went on to develop a model of TQC that consists of the following five elements:

1. Customer obsession
2. The planning process
3. The improvement cycle
4. Daily process management
5. Employee participation

In the remainder of this text, we will take an approach to TQC that encompasses these five elements; we will also go beyond these elements and discuss how to manage and measure a successful TQC effort.

Finally, given some of the data we reviewed earlier, total quality control has to be an integral part of a company's business strategy. This is the key to survival and growth in the 1990s—there is no alternative.

References

1. Buzzel, Robert D., and Bradley Gale. *The PIMS Principles: Linking Strategy to Performance.* New York: The Free Press, 1987. In this book, Buzzel and Gale provide some very convincing data that relates quality with profitability. The relationship is summarized in our Fig. 1.2. To develop this relationship they used a database of about 200 strategic business units (SBUs) from a total database of about 3000 SBUs. The relative perceived quality axis in Fig. 1.2 was determined by getting groups of managers to identify nonprice product attributes that affected customer buying decisions. From these a quality score was constructed.
2. Kano, Noriaki. *Trends of Financial Performance Measures of the Companies Who Won the Deming Prize.* Tokyo: JUSE, 1987. This data was taken from a paper published by JUSE in Japan, 1987. The paper was translated from the Japanese by Katsu Yoshimoto of Yokogawa Hewlett-Packard Company. The database includes 9 Deming Prize companies from the 1960s, an additional 8 in the 1970s, and an additional 17 in the 1980s. The industries from which these companies come include chemical, mechanical, electronics, automobile, and construction.
3. Kano, Noriaki, et al., "Attractive Q vs Must Be Q." *Hinshitsu (Quality)*, vol. 14, no. 2, 1984, pp. 39–48.
4. Feigenbaum, A. V. *Total Quality Control.* Singapore: McGraw-Hill, 1986.
5. Taken from the author's personal notes gathered at an AOTS seminar, presided by Kaoru Ishikawa, at Kuala Lumpur in 1987.
6. Ishikawa, Kaoru. *What Is Total Quality Control?* Englewood Cliffs, N.J.: Prentice-Hall, 1985.

Chapter 2

Customer Obsession

Satisfying customers is the only reason we're in business.
JOHN YOUNG, CEO,
HEWLETT-PACKARD COMPANY

The customer is king.
ANONYMOUS—UNITED STATES

The honorable customer is God.
ANONYMOUS—JAPAN

Overview

In most companies and cultures the customer is considered very important. But how far would anyone go to help the customer? Read the following news article:*

Japan's Got Us Beat in the Service Department, Too

My husband and I bought one souvenir the last time we were in Tokyo—a Sony compact disk player. The transaction took seven minutes at the Odakyu Department Store, including time to find the right department and to wait while the salesman filled out a second charge slip after misspelling my husband's name on the first.

My in-laws, who were our hosts in the outlying city of Sagamihara, were eager to see their son's purchase, so he opened the box for them the next morning. But when he tried to demonstrate the player, it wouldn't work. We peered inside. It had no innards! My husband used the time until the Odakyu would open at 10 to practice for the rare opportunity in that country to wax indignant. But at a minute to 10 he was pre-empted by the store ringing us.

---

*By Hilary Hinds Kitasei. Extracted with permission from the *Wall Street Journal*, July 30, 1985.

My mother-in-law took the call, and had to hold the receiver away from her ear against the barrage of Japanese honorifics. Odakyu's vice president was on his way over with a new disk player.

A taxi pulled up 50 minutes later and spilled out the vice president and a junior employee who was laden with packages and a clipboard. In the entrance hall the two men bowed vigorously. The younger man was still bobbing as he read from a log that recorded the progress of their efforts to rectify their mistake, beginning at 4:32 P.M. the day before when the salesclerk alerted the store's security guards to stop my husband at the door. When that didn't work, the clerk turned to his supervisor, who turned to his supervisor, until a SWAT team leading all the way to the vice president was in place to work on the only clues, a name and an American Express card number. Remembering that the customer had asked him about using the disk player in the U.S., the clerk called 32 hotels in and around Tokyo to ask if a Mr. Kitasei was registered. When that turned up nothing, the Odakyu commandeered a staff member to stay until 9 P.M. to call American Express headquarters in New York. American Express gave him our New York telephone number. It was after 11 when he reached my parents, who were staying at our apartment. My mother gave him my in-law's telephone number. The younger man looked up from his clipboard and gave us, in addition to the new $280 disk player, a set of towels, a box of cakes, and a Chopin disk. Three minutes after this exhausted pair had arrived they were climbing back into the waiting cab. The vice president suddenly dashed back. He had forgotten to apologize for my husband having to wait while the salesman had rewritten the charge slip, but he hoped we understood that it had been the young man's first day.

An incredible story, but true. Surely this is what any customer wants—to have excellent service and products, to be treated well, and to purchase more from a customer-obsessed company. And yet, how few are the managers, companies, or organizations that show this type of customer obsession.

In today's environment, as competition increases and as more sophisticated goods become commodities, the main distinguishing factor becomes quality—the quality of the product or service. In most cases this means the product must be of high quality and be provided via excellent service, and if the customer does not get this, you will lose the customer. In small mom-and-pop-managed or family-run stores you will still find excellent service. Many small organizations or young companies are nimble, quick on their feet, giving the customer what he or she wants—a new product, a new service, or both. But what can larger organizations do to avoid the bureaucracy, complacency, and impersonal attitudes that may eventually cause them to lose touch with the customer?

We will offer a few methodologies that will help ensure a strong customer focus—and preferably a customer obsession. These, together with a well-educated and trained work force, will help you get closer to your customers—to hear their voices—and to meet their needs.

Customer Obsession via a Systematic Approach

Any organization that is serious about quality and customers must take a systematic approach to ensure a customer-obsessed work force. An educated work force is a good start, but it must also have the right tools. Below is a short list of several tools or activities that will foster customer obsession. We have separated our list into reactive and proactive activities.

Reactive activities
1. A system to manage and resolve customer complaints.
2. Customer satisfaction surveys and follow-up corrective action.
3. Collection of product and service failure data, analysis, and follow-up corrective action.

Proactive activities
1. Competitive benchmarking: learning to compete better from "world class" companies.
2. Capturing the customer's voice or needs for new product and services via a systematic process.
3. Focus groups: meetings with customers to understand and extract their views and needs. This is a subset of capturing the customer's voice.

The reactive approach is necessary to understand and resolve challenges and problems arising from current products and services. The proactive approach is essential to help influence and create new products and services.

We will discuss two items from each category: a customer complaint management system, customer surveys, competitive benchmarking, and capturing the customer's voice. We will conclude with a discussion on meeting and exceeding customer needs.

Customer Complaint and Feedback Management System

> One of the strongest signs of a bad or declining relationship is the absence of complaints from the customer. Nobody is ever that satisfied, especially not over an extended period of time. The customer is either not being candid or not being contacted.—*Theodore Levitt*

Have you ever lost your luggage on an airplane? Or had a problem with your personal computer or its software? Or had terrible service at your bank? Did you complain and how was your complaint handled? Was your complaint handled quickly and effectively? If, for example, you lost your luggage and the corrective action by the airline was sloppy and tardy with little sympathy from the airline, you will probably never fly this airline again.

In lean and mean America, this seems to be the trend: profits decline, costs are cut, people are fired, resulting in deteriorating service, and customers go elsewhere—to a company that understands that service and customer satisfaction are essential for success. This pattern is very visible in the airline, banking, and retail sectors, but it is also common in all other sectors.

When a customer is upset, you may get a complaint coupled with a demand for compensation. On other occasions, customers may give you feedback because they genuinely want you to improve your product or service. *Each customer complaint should be treated like an unpolished gem; a gem that needs to be captured, examined, and polished.* Your company or organization can only be richer and wiser as it collects and polishes each of these gems of insight and wisdom.

Having a system that captures these gems of wisdom and polishes them is extremely important. If you don't, the wronged or unhappy customer may not return to you again. Consider the following information released by the United States Office of Consumer Affairs, on March 31, 1986:

The High Cost of Losing a Customer

- In the average business, for every customer who bothers to complain, there are 26 others who remain silent.
- The average "wronged" customer will tell 8 to 16 people. (Over 10 percent tell more than 20 people.)
- 91 percent of unhappy customers will never purchase goods or services from you again.
- If you make an effort to remedy customers' complaints, 82 to 95 percent of them will stay on with you.
- It costs 5 times as much to attract a new customer as it costs to keep an old one.

The most shocking statistic is that 91 percent of unhappy customers will *never* purchase goods or services from you again. The following box contains an interesting memo from Kent Stockwell of Hewlett-Packard Company, entitled "A Customer Lost Is a Customer Lost," that supports this statistic.* Notice that customers have very long memories—they remember both the good or bad that you do for a long time.

The need for a system to manage customer complaints

Every company or organization is always receiving customer complaints. The difference between each organization is the frequency and intensity of the complaints. Some organizations will ignore the complaints, while others with a well-informed and educated work force will find that complaints are attended to. But this alone is insufficient. The better organization must keep a record of these complaints—the frequency, the intensity, the location, and so on.

To track lost luggage, an airline's management would need to know where, when, and how often luggage was being lost and if it was happening in one airport or in every airport. Only with such data could the airline move away from solving lost luggage on an ad hoc basis to repairing or correcting the entire system—in order to eliminate lost luggage.

A Customer Lost Is a Customer Lost

I called an HP calculator customer in Los Angeles to ask him some survey questions, following up on our recent customer satisfaction survey. This person was a very busy civil engineer who runs his own consulting company. Each time I called, I was asked to call back at a specified time. The third time I called, I asked to wait, and I finally got through.

After going through the questions for 10 minutes, I asked the final question: "Is there anything else you would like to say about HP calculator products and service?" This customer said, "No, your calculators are great but your computers have a problem."

Then he proceeded to talk for another 10 minutes (this very busy person) about his negative experiences with an HP desktop computer that he had purchased 12 years ago. The computer he had purchased was delivered late, with no explanation. And he was unable to get help from the sales office regarding missing cables and questions he had about the manuals and the system. Finally he called the factory in Loveland, where he got help and found out about training classes he could arrange through the sales office.

Today, 12 years later, he still remembers the story in detail; most important, this technical computer user owns ABC equipment, not HP. Because of some minor product problems and some major service problems, this former HP customer has not looked to HP for his technical computer needs for over 12 years.

---

*Printed by permission of Kent Stockwell, Quality Manager, Hewlett-Packard. *Note*: The name of the competing company has been disguised.)

Hence, it is vital for your work force to record most, if not all, complaints. This must be followed with resolution of the complaints and elimination of the root cause of each complaint.

Customer complaint and feedback system

Figure 2.1 shows a flowchart of a customer complaint and feedback system. The flowchart is simple and self-explanatory. The key points include the following:

Complaints are collected from all sources. Complaints from letters, phone calls, meetings, and verbal inputs are collected routinely.

Data collection via a customer complaint and feedback form. Data collection and transmission at the local entity is done via a customer complaint and feedback (CC&F) form (see Fig. 2.2). This same form can go to the central coordinator, who will use a formal corrective action request (CAR) (see Fig. 2.3). More on this later.

Quick resolution. The complaint is resolved as quickly as possible, and the customer is contacted and informed.

All customers get a response. This can be a thank-you note or a solution for a complaint.

Issues are resolved locally. For a locally generated issue (that is, generated at the site which collected the complaint), the issue is resolved locally.

Issues are resolved by the organization. For an issue beyond the control of the local entity (that is, generated elsewhere in the company), the information is sent to a central coordinator or analyst. This coordinator will further analyze the issue and propose a solution to the process owner in the organization.

Resolution of systematic issues. On a regular basis, data is analyzed and systemic issues are identified, resolved, and eliminated. Not doing so can have an impact on future purchases by current and potential customers. It can result in lost business and reduced market share. Here again you can use the CAR form.

Identify and monitor performance measures. Regularly review performance measures such as number of complaints each month, time to resolve complaints, Pareto diagrams of complaints, and so on. Keeping track of these and setting appropriate targets will help provide proper management of the system and give an indication of its health. Targets should be set only after sufficient data on the current status is available. For example, if it takes an average of 20 days to resolve complaints, a target of 10 days can be set, and this target can

Customer Obsession   19

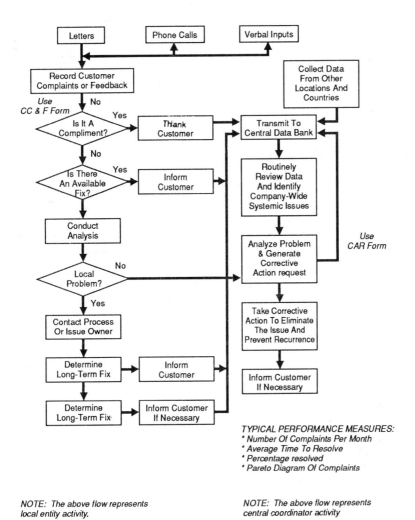

Figure 2.1 Flowchart of a customer complaint and feedback system.

be gradually reduced over months and years. Another example of appropriate targets would be the elimination or gradual reduction of certain categories in the Pareto diagram of complaints. We will discuss this point, in greater detail, later in this section.

Recording customer complaint data. Customer data can be recorded either on a paper form such as a CC&F form as shown in Fig. 2.2, or the data can be keyed into a computer-based form. It will not be easy to record the issue on paper or in a computer database. A busy sales rep-

```
┌─────────────────────────────────────────────────────────┐
│ Customer Complaint & Feedback System (CCF)              │
│                               ┌─────────────────────────┤
│   Customer Name:_____ │ RECIPIENT DATA:         │
│                               │ Name_____         │
│   Address_____ │ Date_____         │
│                               │ Telephone No_____ │
│   Contact Number:_____ │                         │
│                               │ Contact Type:           │
│   Product Or Service Impacted:│ ☐ Letter                │
│                               │ ☐ Telephone             │
│   Customer's Comments:        │ ☐ Personal Contact      │
│   (In Customer's Own Words)___│ ☐ Others_____         │
│   _____│                         │
├─────────────────────────────────────────────────────────┤
│   Additional Comments:_____               │
│   _____           │
│   _____           │
├─────────────────────────────────────────────────────────┤
│   CUSTOMER REQUESTED ACTION:                            │
│   ☐ Compliment  ☐ Compensation Required  ☐ Formal Reply │
│                                                         │
│   ☐ Information     ☐ Replacement   ☐ Others_____    │
│   (No Action Needed)  Or Cancellation                   │
├─────────────────────────────────────────────────────────┤
│   DEGREE OF SERIOUSNESS:                                │
│   ☐Normal  ☐Safety  ☐Requires Escalation                │
│                                                         │
│       ANALYSIS AND PROPOSED                             │
│       CORRECTIVE ACTION:                                │
│                                                         │
│       _____                           │
│       _____                           │
│                                                         │
│    Can the problem be resolved                          │
│       locally?                                          │
│                                                         │
│       _____                           │
└─────────────────────────────────────────────────────────┘
```

Figure 2.2 Typical form for recording customer complaints.

resentative or manager may at best be able to phone in the customer's complaint. A busy counter clerk may resolve the problem immediately but may only have time to scribble a few words. Other very analytical employees may insist on writing down a detailed history of what happened.

Hence a flexible system is needed—one which provides easy ways of collecting inputs. For example, inputs can be taken via telephone to a central voice mailbox; directly to a complaint specialist, via an electronic mail system; or on the recommended form shown in Fig. 2.2. As the system matures and gains credibility, recording the data will become easier, but employees will always appreciate having several options.

Data collection via a telephone voice mailbox. If you decide to experiment with a telephone voice mailbox, a cue card can be provided to all employees. This little card can be used during discussions with customers and to phone in the problem to the customer complaints voice mailbox. An example of a cue card is given in Table 2.1.

Corrective action requests. An issue may be resolved locally depending on the situation or type of complaint. But when a class or companywide problem is being resolved, it will be sent to a central coordinator for further analysis. In such cases it will be necessary to ensure that the problem is resolved and eliminated throughout the organization. If this process is done properly, the problem should never recur in the organization. One way of proposing a companywide fix is by sending the proposal, via a CAR form, to the process or problem owner. This, like the CC&F form, can be transmitted electronically. But initially, you should try this on a manual system.

A hypothetical example of a completed CAR form is shown in Fig. 2.3. This, when completed, will be sent to the process owner within the organization. This example is based on an actual problem a major automobile manufacturer had. Unfortunately, this manufacturer took about 4 years to correct the problem. That is approximately the product life cycle of some cars and possibly more than the average develop-

TABLE 2.1 Cue Card for Customer Complaint and Feedback System

Cue card for voice mail
Phone number of telephone voice mailbox: XXX-YYYY
Please have the following information available: Your name, department, and phone number Customer name, title, address, and phone number Customer's actual words describing the problem What the customer wants Any additional information about the problem Model/product number, serial number, sales order number, or service affected Action already taken Specific help needed, from who, when, or how Any other comments

CORRECTIVE ACTION REQUEST (CAR)		Reference No:	
PRODUCT/PROCESS OWNER:	ABX	MAIL TO:	ABX Operations Mgr

Item	Windshield Wiper		
Product:	Xetta Automobile	Category	All 1984 Models With R.H.Drive
Requested by:	CC & F Analyst	Date	April, 1984

SERIOUSNESS: Is This a Class Problem: Yes [X] No [ ]   Dealers grumbling and calling in
What is the current Impact? Unhappy customers, sales are slowing down. Estimated loss in Sales:$XXXXX.
What is the future Impact? Loss of reputation, loss of future sales, $ impact unknown.

SUMMARY OF REQUEST: (ADD MORE DETAILS, WHERE APPROPRIATE)

We have received over 350 complaints via the CC&F system from several countries. (See attached data). For all 1984 models shipped to right-hand drive countries, the windshield wiper does NOT provide sufficient clear vision during rain.

ANALYSIS:

Our Analysis, and that of dealers and customers is as follows: The windshield wiper is in the position required for left hand drive cars (Continental Europe). But the cars are shipped to right-hand drive countries with right-hand steering. However position of wipers was not changed. As a result, during rain, the vision is clearer on the front passenger's side, but poor on the driver's side.

PROPOSAL: (ADD DETAILED PROPOSAL, WHERE APPROPRIATE)

The solution is obvious. We need to retro-fit wipers on current cars to the correct position. And, ensure future production is done correctly.

COMMENTS BY PRODUCT/PROCESS OWNER (Including $ Impact):

This product was designed correctly. However the manufactuing documentation missed this point. We have done the update.

We estimate retrofit will cost $YYY. And tooling costs of $NNN.

ACTION:
Short-Term Fix   Immediate retrofit kit to dealers
Long-Term Fix: Stop R.H.Drive production;until documentation and tooling is redone. Estimate is 3 weeks.

Figure 2.3 A completed corrective action request form.

ment time for a new car at Toyota Motor Company. Obviously, the manufacturer we quote did not have a good system to manage customer complaints.

Promoting and facilitating the system

Promotion activity. A customer complaint and feedback system will require constant nurturing and promotion, especially in the first few years after it is launched. Here is a list of methods to explain the system and its benefits:

Promote during new employee training. New employees are typically most receptive to such a system. Their participation should be requested.

Promote at monthly department meetings. This needs to be done on a very regular basis and reduced, but not eliminated, as the system becomes accepted. At these sessions, encourage participation, review issues, and convey success stories of resolved customer complaints.

Encourage managers to use the system. Managers should use the system and become role models for their employees. Their role is crucial as they can help to break down resistance to change. For example, a general manager at a hotel can talk to hotel guests and understand their gripes or compliments; or a general manager at a large corporation can call several customers and find out how the new system is performing at the customer's site. The data obtained can be fed into the formal customer complaint and feedback system.

Facilitation of the process. The organization's quality managers must help to manage the entire process. The specific activities for the quality managers and their staff will include the following:

Promote use of the system. This is discussed above.

At the organization level. The organization's president or director of quality can oversee the entire system and also provide a central coordinator (see the flow chart in Fig. 2.1).

At each entity within the organization. The entity's quality manager can manage the customer complaint and feedback system and help resolve locally generated problems and issues (see Fig. 2.1). The requisite activities include:

- Record all complaints and monitor their ongoing status.
- Analyze the root cause of the complaint: why it occurred and how to avoid recurrence in the future.
- Ensure proper routing of customer complaints to the responsible process or issue owner.
- Request follow-up if an employee, manager, or entity is lax.
- Acknowledge receipt of the complaint and specify the corrective action taken to the individual who reported it.
- List all complaints and corrective action as part of a monthly quality report. Document the Pareto diagram of complaints.

- Review complaints, especially outstanding complaints, at the regular entity TQC steering committee meeting. Refer to the flowchart and discussion provided in the quality assurance system in Chap. 5, "Daily Process Management."
- Escalate a complaint to senior management if corrective action or progress is deemed unacceptable. Recommend appropriate action.
- When necessary, ensure that customers are informed of progress of their complaint.

Extending the system to customers. The customer complaint and feedback system can be extended to external customers—but only after it's working well inside the company.

Customers can dial a hot-line telephone number. The number should not be a telephone voice mailbox but one attended by a warm body. The quality department can manage this service. Initially, the service can be extended to preferred customers; later it can be provided to all customers.

The benefits of extending the system to customers include the following:

- It will serve as a safety net to catch customer complaints that are managed poorly by the normal system.
- It will convince customers that you care.
- It will give you a competitive advantage.

Even the best company will benefit by doing this; and we know of several companies that have done this successfully. In the United States, for example, Marriot Corporation and American Express have installed hot lines to collect customer complaints. In Britain, British Airways has installed video booths at London's Heathrow Airport to collect customer complaints—this is a very innovative way to collect "hot" customer inputs, immediately after the trip. And yet, as far as we know, nobody else has imitated British Airways.

When the system is extended to customers, employees should be encouraged to go out of their way to help customers. A way of doing this is to empower employees to make decisions that cost, say, $10 to $100 or to allow product replacements for dissatisfied customers. The impact of such a decision can be electrifying to customers and employees. The result can be a quantum jump in customer satisfaction. We suggest you try this, after the customer complaint and management system is running smoothly.

Customer Obsession 25

What Can You Expect If You Start a Complaint System?

The various categories of complaints that you can expect, in a design, manufacturing, and sales organization, are discussed in the next section. Here we give a random list of complaints from the automobile, computer, and instruments industries. They range from thoughtful to funny. All represent problems that need resolution; and all have been sanitized.

- Your component failure rate is 150 parts per million (or 0.015 percent). You need to improve it! (*Comment:* Imagine the response if the failure rate was as high as in the complaint that follows. This is a good example of a customer and supplier working to get to zero defects.)
- About 15 percent of your products are defective on arrival!
- Your parts and service are too expensive, and the service is slow! Therefore I go elsewhere for support.
- You sold me a car with the windshield wipers fixed incorrectly and the dealer says it's a manufacturing defect. Now, whenever it rains, I cannot see clearly. (*Comment:* This is discussed elsewhere in the section on corrective action requests.)
- The instrument arrived with the glass tube broken. This has happened before. Please replace or repair.
- Why bother to survey again? You never did anything to help me since the last survey! (*Comment:* Scrawled across a blank survey form.)
- You promised a delivery date of *xxxx*. Now you say you can't make it. Let it be known that if you don't, there will never be any more business for you!
- The invoice is illegible! I am not going to pay you anything until you send me a proper invoice.
- Using my car's turn signal is like breaking a chicken's leg!
- Take back your product! (*Comment:* Scrawled across a blank customer survey form)
- You shipped a product and I never ordered anything. I am not going to open the box!

Trends in a customer complaint and feedback system

Will complaints occur—always? The answer is Yes. Humans are not perfect, our knowledge is incomplete. Hence errors will occur and complaints will occur. But we should manage complaints by doing the following: Document, resolve, control, minimize recurrence, and reduce the intensity of complaints. This will result in a change in the trends of complaints.

Trends of complaints in a design, manufacturing, and sales company. Consider a design, manufacturing, and sales company—large or small. When it initially starts to record customer complaints, it will find that the majority of the issues relate to the interface between the company and the customer. Interface issues are those involving employees and the customer: issues such as administrative problems, rude or careless employees, cosmetic problems (dirty or damaged products), miserable after-sales service, and delivery problems. As these interface issues are eliminated, product (or service) design issues become paramount. In addition, new categories of complaints may appear as the company's business strategy changes.

Look at Fig. 2.4. It is constructed from data gathered after implementing a customer complaint and feedback system in a country operation of a large multinational manufacturing and sales company.

In the initial stages, interface issues, such as product delivery (late delivery, damaged items, etc.) and product support issues (after-sales support, repair quality, costs, etc.), abounded. But as awareness of these issues increased, the systemic root causes of these issues were gradually eliminated. With time the paramount customer issue became one of product design.

As shown in Fig. 2.4, after several years, the percentage of product design issues increased. Typically the number of product design issues will also increase. This is not because there are more design problems, but because the company is now more sophisticated—proper analysis, training, and new methods have eliminated the interface issues, while allowing the detection and understanding of product design problems facing the customer. The company can now focus on reducing design problems via several tools that are discussed in Chap. 5, on "Daily Process Management."

In Fig. 2.4, the second issue recorded several years later turned out to be problems caused by the company's dealers and distributors, who

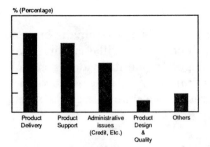
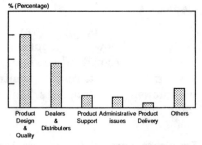

Figure 2.4 Pareto diagrams of trends in a customer complaint and feedback system at a design, manufacturing, and sales company.

sell this company's products. This category did not exist earlier because of the nature of the business and product strategy.

Trends of complaints in the service industry. The same situation would occur at a service organization, such as an airline or a bank. For example, at an airline, initially complaints about interface issues will abound. From our experience, these interface issues will include telephone courtesy, counter service, baggage handling, and passenger management on busy flights. Later, as these issues are resolved (and this is the case with a well-managed and mature airline), complaints about the design and sophistication of services will increase, such as quality and variety of meals, internal decor, design of seats, and special handling of baggage on first and business class flights.

Summary: Customer complaint and feedback system

Errors will occur; hence every company or organization will get customer complaints. These problems must be monitored, managed, and eliminated—with the help of a formal customer complaint and feedback system. Elimination or reduction of many categories of complaints can occur if systematic root causes are removed.

As the systemic root causes of the complaints are removed, the trend of the complaints will change. Complaints that pertain to design of products and services will become evident and increase, especially in a design and manufacturing company.

Customer Satisfaction Surveys

> Talking to customers tends to counteract the most self-destructive habit of great corporations . . . that of talking to themselves.—*John Brooks,* Telephone—The First 100 Years.

There are several kinds of customer satisfaction surveys. We will discuss just a few here:

A postpurchase survey. This is conducted after a product is purchased by the customer. This can be done via a response card that comes with the product which the customer completes and returns to the manufacturer or distributor of the product. This can provide a wealth of information about the product—its quality, ease of use, the quality of the product handbook or manual, the ease of doing business with the company, and so on.

A postinstallation survey. This is conducted after a product is delivered and installed at the customer's site. This survey can be done via mail or, preferably, via a telephone survey. It is for a sophisticated product that requires installation by the manufacturer or distributor. Some topics to check for in the survey are delivery commitments, courtesy of the staff, speed of installation, and the quality of the solution.

A customer satisfaction survey. This survey measures the customer satisfaction level with a company's products and services. There are numerous specialist companies who do third-party surveys. For example in the United States, Datapro conducts surveys on customer satisfaction for computer companies, and John D. Powers does the same for the automobile industry. In addition to these third-party surveys, which are very helpful in competitive analysis, a survey which measures your customers' satisfaction with your company is required. Many large corporations such as Hewlett-Packard, Florida Power & Light, and Xerox, conduct such surveys in the United States. We will briefly discuss this type of survey.

Conducting customer satisfaction surveys

Such a survey should measure customer satisfaction with the various attributes of your products and services. For example, if you are a design, manufacturing, and sales organization, you would measure satisfaction in areas such as:

- Marketing programs
- The sales force's interaction with customers
- Administrative services (credit, invoicing, etc.)
- Product delivery and installation
- Product education and training
- Product documentation and information
- After-sales service and support
- Product quality
- Effectiveness and value of the solution provided
- Overall cost of ownership (from purchase to use)
- Ease of doing business with the company

Each of the attributes would break down into a group of questions. For example, the sales force's interactions with customers would include the following:

- Availability of sales representatives
- Technical knowledge of sales representative
- Understanding the customer's real needs
- Consulting skills of the sales representative
- Predicting the customer's future needs

All questions can be measured on a scale of 1 to 5 or 1 to 10. Consistency of measurement is important to allow trends to be observed over several years. In addition to measuring satisfaction, each question or group of questions should be measured for importance.

Customers who complete the survey should be encouraged to provide comments for each item, as well as random comments. We recommend doing a survey of a random sample of customers. But for the sample you should try to get close to a 100 percent response. If the response rate is low, research on nonrespondents must be done.

Setting priorities for issues and customer comments

The issues that customers face, and which must be resolved, can be ranked in several ways:

In order of survey score. You will resolve the items with the lowest scores—that is, lowest satisfaction.

Selected from a scattergram. You can construct a scattergram showing both score and importance of each attribute. This can be done if you also measure each question or attribute (group of questions) by its importance to the customer. Hence, the customer may give you a low satisfaction score for administrative services but may also rank them low in importance. Your scattergram may look like the one in Fig. 2.5. For clarity, we show only three attributes.

Look at the scattergram: both administrative services and delivery score low in satisfaction. But delivery has a higher importance level. If resources were limited, you would work only on delivery. If resources were plentiful, you would do both. At the same time product quality, although important, is providing high satisfaction—no additional resources need to be diverted to this attribute. But the current quality must be maintained. This is where daily process management becomes important. There is more on this in Chap. 5.

Selected from a Pareto diagram. A Pareto diagram of customer comments can be prepared. Many customer comments will require

Figure 2.5 Scattergram of satisfaction versus importance.

further contact with the customer. But the numerous customer comments can be converted to quantitative data. How would you do that? We recommend the use of the KJ method. If KJ is unfamiliar to you, refer to the section "Seven New Quality Control Tools" in App. 2. This methodology will allow you to group verbal data with affinity. For example, complaints on delivery and installation or about poor dealers and distributors can be grouped. The resultant groups can be displayed on a Pareto diagram. This will give you another list of priorities.

Customer satisfaction model

The surveys we mentioned earlier can be used to prepare a customer satisfaction model for each product. Figure. 2.6 shows a model listing several elements of a customer's satisfaction for a sophisticated product, like a printer, computer, railway engine, or airplane. Notice that the elements go beyond the traditional reliability of a product. Instead, we measure a full spectrum of product attributes or elements, such as those shown in Table 2.2. Each element can be segmented into subelements, with appropriate targets. Examples are given in the table for reliability, ease of use, and product availability and delivery. Other elements are listed without elaboration. Such data will have to be collected from several sources.

Targets can be set for the various elements or performance measures, after appropriate data analysis. These targets then become business fundamentals or daily management items for the products. They must be reviewed regularly and achieved. Refer to Chap. 3, "The

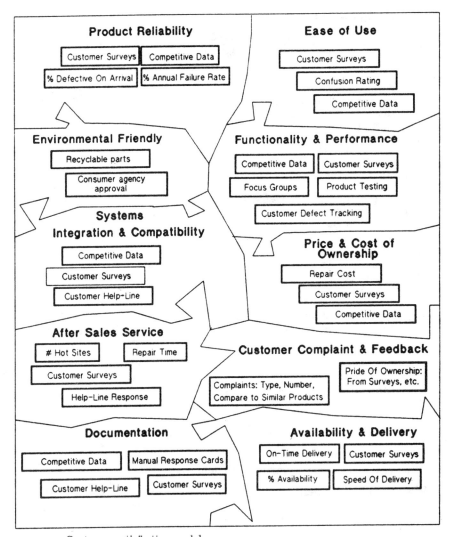

Figure 2.6 Customer satisfaction model.

Planning Process," for a discussion of Daily Management, or Business Fundamentals.

Who owns and manages the customer satisfaction model? The entire company does, but the company quality department can facilitate this activity and ensure that targets are set, results measured, and corrective action taken. This customer satisfaction model, if managed well, can become a powerful tool to collate, measure, and achieve customer satisfaction at any entity that designs, manufactures, and sells products.

TABLE 2.2 Customer Satisfaction Performance Measures

Product: Essex 400X, 420X, 480XE

Item (to be measured every 6 to 12 months)	Target	Actual	Owner
1. Reliability			
Annual failure rate (%)	1.0	1.1	Manufacturing
Defective on arrival (%)	0.2	0.1	Manufacturing
Customer's reliability rating	1	1	Quality Assurance
Competitive reliability rating	1	2	Quality Assurance
2. Ease of use			
Customer rating of confusion factor*	>8.5	8.9	Research & Development
Competitive rating*	>8.5	8.6	Research & Development
3. Availability to dealers and retailers			
Product and accessory availability (%)	>95	93	Distribution
On-time delivery (%)	>97	88	Distribution
Speed of delivery (after order)(hr)	29	27	Distribution
4. Documentation			
5. Price and cost of ownership			
6. After-sales service			
7. Systems integration or compatibility with other products			
8. Functionality and performance			
9. Customer complaints and feedback			
10. Environmental friendly product (an issue for the 1990s)			

* Measured on a scale of 1 to 10.

NOTE: This list is derived from ideas generated by Joseph Juran's discussion on product fitness for use.[1]

Other types of customer satisfaction data

We would like to mention one other kind of customer satisfaction data. This is a list that your customer keeps. For example, many companies keep a regular score card of each supplier's performance. You may want to ask your major customers to tell you what they think of you. If they do not have such data, in your next visit to a major account you should suggest they collect it. Since this could be a broad brush assessment of your performance, you could start the discussion by asking for information in the following areas of your products and services:

- Technology
- Reliability and quality
- Effectiveness and value of the product or service
- Cost of ownership (from purchase to use)

- Ease of doing business and responsiveness
- Delivery

This is a shortened and modified list of what we show in Fig. 2.6. Typically this kind of information can be rated on a 0- to 5-point scale and displayed on a radar chart, similar to the one we will discuss in Chap. 8, "Conducting TQC Audits or Reviews." Comparisons to competitors can also be made on the chart.

Ensuring follow-up to customer satisfaction surveys

The most crucial part of a customer satisfaction survey is to take corrective action after the data is collected and analyzed. In App. 1 we document an example of corrective action taken after a customer satisfaction survey is conducted.

Follow-up is the responsibility of management. The Xerox Company in the United States has an extremely customer-obsessed management that tracks customer satisfaction diligently. They tie part of management compensation to customer satisfaction as measured in a survey. Many managers would disagree with the Xerox method—we certainly do. The better way is to plan for it, by setting the correct expectations. This will be discussed in the next chapter, "Planning Process." Nevertheless, Xerox has had success in increasing customer satisfaction. It recently won the coveted Malcolm Baldrige Award in the United States. It is also gradually increasing its market share after years of decline.

Competitive Benchmarking

Competitive benchmarking has been around for thousands of years. More recently it was popularized by Robert C. Camp in his book *Competitive Benchmarking*.[2] Competitive benchmarking is one of the strategies employed by the Xerox Company to make itself more competitive by requiring "world class" processes in its operations. Robert Camp's book is a result of this effort.

Competitive benchmarking is an important process that allows us to measure our products, services, and processes against the toughest—preferably world class—or leading edge companies. This methodology forces us to have an external view that can catalyze internal change in our company or organization, and it helps us to be more efficient and competitive.

Purpose of benchmarking

Before we continue, we wish to stress that the purpose of competitive benchmarking is not to conduct industrial espionage or to do anything unethical. Rather, the purpose is to look at processes used by other leading companies and to use that learning to improve our processes. This will provide us with competitive advantages for a short duration until another company arrives with a better way or new benchmark—because the improvement process is continuous. At the same time, we can expect other competitors and leading companies to emulate the best companies.

What can be benchmarked?

Almost everything can be benchmarked—both products and services. We will give examples of both. A partial list for a design and manufacturing environment is shown below, but the same concept can be applied to the service industry, as shown in our examples.

- Manufacturing methodologies
    - Inventory control and management
    - Warehousing and delivery services (more on this later)
    - Production technologies
    - Automation techniques
- Product performance (more on this later)
- Quality systems
- Product development process
    - Understanding and collecting customer needs
    - Design tools and technologies
    - Process of quick design and introduction to market
- Marketing techniques

Benchmarking process

Shown below is a suggested benchmarking process. This is fairly generic and applies to benchmarking processes or product performance. We will then elaborate on some of the steps, using a product and process example.

1. Identify subject (product or process) to be benchmarked and set up a team with a leader.
2. Identify world class or leading edge companies that have a similar product or process.
3. Prepare data collection strategy.
   - Approach a world class or leading edge company
   - Obtain a competitive product
   - Obtain comparative data of our own product or process
4. Collect and analyze data after visiting the leading edge company or examining a competitive product.
5. Determine performance gaps between the benchmarked subject and our product or process.
6. Discuss and determine performance targets to be achieved with the team. Prepare list of recommendations and get management approval to implement them.
7. Prepare an implementation plan for the recommendations.
8. Implement the improvement plan.
9. Monitor progress and ensure successful implementation.
10. Review and recalibrate the benchmarked subject.

Most of the steps are quite straightforward, but let us review some of them.

Look at step 3. We must look at world class or leading edge companies. This is an important step because the world class process we are trying to emulate may not be from a competitor. If we are in the computer industry, the best purchasing or distribution process may be in the supermarket sector. For a computer product, the best manufacturing technology may be in another industry, such as cameras. Therefore it is important that we look at the overall best external practice and not just that within our industry or competition.

Next, look at step 4—collect and analyze data. The actual data required in this step will vary; it will depend on the type of product or service being analyzed. The review team will need to meet and discuss these requirements in great detail before setting forth on this step. If we were analyzing an electrical or electronic product, here is what we would look for:

1. Delivery and cost information
   - Availability of product
   - Price of product
   - After-sales maintenance costs
2. User friendliness
   - Simplicity and usefulness of manuals
   - Product packaging quality
   - Ease of use of the product
3. Performance review
   - Speed of operation
   - Feature set
4. Design review
   - New ideas and concepts
   - New technology
   - Style of product
5. Manufacturing methods review
   - Design for manufacturing (ease of assembly)
   - Cost of materials and manufacturing
   - Process technology
6. Environmental testing
   - Radiation and interference generated
   - Environmental friendly attributes, such as packaging, noise generated, and energy costs
7. Stress testing
   - Electrical characteristics
   - Vulnerability to shock and vibrations
   - Vulnerability to temperature and humidity

If we were benchmarking a process, we would arrange to visit a friendly company or organization with a world class process. For a distribution process, here is what we would look for:

1. Products distributed
   - Markets and customers served
   - Types of products
   - Volumes of products
   - Shipment dollars
2. Warehouse facilities
   - Locations: local and remote
   - Size
   - Security features
   - Automation methods
   - Computer information systems
3. Shipments
   - Packaging methods
   - Shipping methods
   - Shipping costs
   - Shipment transit times
4. Order processing
   - Methods of ordering by customers
   - Ease of ordering by customers
   - Order processing costs
5. Use of third-party or independent distributors
   - Distributor used
   - Services provided
   - Costs and savings incurred

6. Efficiency and quality of services
   - Costs of operation: wages and overheads
   - Weight, volume, and dollars shipped per employee
   - Inventory turns per year
   - Quality of shipments: damage, delays, failure to meet commitments, etc.
   - Ability of coordinating several orders into one shipment
   - Supplier response time: average time taken to ship, after the order is received
   - Line order fill rate: percentage of time items are in stock when an order is received
   - Turn around time: total time taken after customer orders until delivery to customer's doorstep
   - Customer complaints and feedback information
   - Employee turnover
   - Training and education provided
   - Return on assets

Once this information is gathered, performance gaps of the benchmarked subject can be better understood. The resulting improvement plan can result in a more competitive product or a world class process in our company. The outcome can be higher productivity, lower expenses, more satisfied customers, and increased profits.

Capturing the Customer's Voice and Needs

Before designing and building products, it is important to capture the customer's voice and needs. This information can help us provide the right product for our customers. And yet, numerous manufacturers design and build products that are either never released or do not sell—resulting in canceled projects, wasted resources, higher costs, and often, lower morale.

> **Other Thoughts on Competitive Benchmarking**
>
> Competitive benchmarking analysis, at best, helps you be the equal to the toughest competitor or industry leader. But it will not let you leapfrog a competitor or predict what the future holds.
>
> Was the American automobile industry able to predict the success of the Japanese car industry in America? And was it able to predict that Japanese manufacturers would move upscale into the luxury car market—via Toyota's Lexus and Nissan's Infiniti? Probably not, on both accounts.
>
> A recent article in *The Economist* magazine[3] discusses the need to go beyond competitive benchmarking. The article quotes Mr. Pankaj Ghemawat, an associate professor at the Harvard Business School, as saying that most companies only benchmark price and performance of their product with a competitor or the entire industry. This methodology will fail to spot potential rivals.
>
> One option is to take the approach of NEC. When it plans for the future, NEC (as quoted in the *Harvard Business Review*[4]) concentrates on honing its "core competences"—its key skills and technologies. In planning for the future, NEC decided that its communications, computer, and component businesses would increasingly overlap. It predicted that success in these businesses would require increasing skills in one specific core competence—semiconductor manufacturing. It invested heavily in this technology—in fact, NEC's semiconductor division recently won the coveted Deming Prize in Japan, thus establishing itself as a high-quality, customer-oriented supplier. This strategy has put it in a commanding position to succeed in computers and telecommunication.
>
> The article goes on to give another option by quoting the example of Chaparral Steel, a profitable American steelmaker. This company does not waste time in trying to improve on its existing products. Instead it visits competitors and university research departments with the purpose of detecting trends—this allows it to predict the next successful product.
>
> Finally the article gives an interesting conclusion: The next rival or breakthrough laser printer may not come from Hewlett-Packard or Apple—the major suppliers of laser printers. It could come from Cannon, who makes 80 percent of all laser printer engines.

Why do products succeed in the marketplace?

Why do some products, which were expected to be blockbusters, fail in the marketplace? And yet other products, which were developed at the whims of a chief executive, succeed beyond all expectations—two examples that come to mind are the Hewlett-Packard HP 35, the first electronic scientific calculator and the Sony Walkman, the first personal stereo tape cassette player.

The difference is customer acceptance. But how do we get customer acceptance? The chief executives of Hewlett-Packard and Sony were willing to bet, by gut feel, that what they proposed would be blockbusters, and they were right. Unfortunately, this process is not easily repeatable, and we need to use a more systematic process that minimizes failure. Is there a better way?

Project Sappho study. A study called *Sappho*[5] (Scientific Activity Predictor from Patterns with Heuristic Origins) was conducted to understand successful industrial innovation. The study was published by the University of Sussex in 1972. The title of the report was "Success and Failure in Industrial Innovation."

The report describes the key factors for success when designing new products and new processes (in the chemical and similar industries). The factors, in descending statistical order of importance, are:

1. Successful innovators were seen to have much better understanding of user needs.
2. Successful innovators pay much more attention to marketing.
3. Successful innovators perform their development work more efficiently than do failures but not necessarily more quickly.
4. Successful innovators make more effective use of outside technology and scientific advice, even though they perform more of the work in-house. They have better contacts with the scientific community, not in general but in the specific area concerned.
5. The responsible individuals in the successful attempts are usually more senior and have greater authority than their counterparts who fail.

Unfortunately this vital research was not well used to increase the success ratio of new products or services by the western industrialized nations like the United States or Great Britain. In addition, the research itself has remained a well-kept secret. But we can clearly see that the most vital ingredient is an understanding of user needs. Let us now look at another study that probes into why products fail.

Why do products fail in the marketplace? A study at Hewlett-Packard came up with many reasons why products fail after release to market. Among them were the following quoted by John Doyle, executive vice-president:[6]

1. Lack of project endorsement by upper management.

2. The better mousetrap that nobody wanted.
3. The "me too product" that meets a competitive brick wall.
4. The technical dog—a product so technically sophisticated that no one understood it.
5. The price crunch—the market wanted a Chevrolet, but we gave them a Cadillac.

This data is not ranked. We do not know the biggest contributors, but items 2, 4, and 5 relate to the customer's needs and perceptions.

Why are products terminated during the design stage? Very often product introductions are canceled because the expectation is that customers will not buy them. In addition, companies have their own reasons for terminating products during the design stage. The result is wasted product development effort. In this area there have been some studies. Let us review one of them.

The Union Carbide Industrial Gases Division (UCIG) did an analysis of wasted product development.[7] They estimated that 12 percent of UCIG's annual development effort was wasted on terminated projects. There were three main reasons for termination:

1. *Customer/market reason.* The project did not meet customer needs, potential customers too few, or alternate products do the same thing and are cheaper.
2. *Business reasons.* Project outside UCIG's domain. That is, the project was not endorsed by upper management.
3. *Technical reasons.* Project failed to achieve technical objective.

The study has similarities with the Hewlett-Packard study. According to UCIG, the first item listed above, customer/market reasons, contributed to the bulk—86 percent—of the wasted development effort. In fact, their analysis defines the root cause of the customer/market problems as follows:[7]

> Many projects are selected and continued without management and the developers having sufficient market intelligence and/or understanding of customer needs and benefits. This results in a needless expenditure of effort on projects for which termination is inevitable.

Wasted product development and product failures will always occur because risks have to be taken, but these must be minimized. This is possible if the reasons are understood; the data from UCIG gives some of the reasons and possible solutions. The solution that UCIG proposed includes the following:

- A written information-gathering plan
- Agreed upon responsibilities and trade-offs
- Milestones and decision points
- Customer feedback for verification and periodic reviews

Summary: Why products succeed in the marketplace. A common thread runs through the three studies that we have quoted: *To have a successful product, we must understand the user's needs.* There will always be a place for the brilliant chief executive or designer who dreams up a blockbuster product. But for most product designs, we will need to rely on a better way—by using a process that attempts to capture user needs. We will discuss this next.

Quality function deployment

A good product definition process is crucial for ensuring that a product or service will meet customer expectations and needs. The product definition process is part of the product development process, which starts with customer needs and ends with delivery to the customer. The same concept applies to a service, where we will need a good service definition process.

The Quality Function Deployment (QFD) methodology is a process that helps to provide better product definition and product development. QFD is a systematic process that helps a design team to define, design, manufacture, and deliver a product or service that meets or exceeds customer needs. The key features of QFD are:

It captures the customer's voice. The customer's voice is captured in order to define product or service specifications.

It ensures strong cross-functional teamwork. Teamwork is ensured between the various functions involved with the design, such as marketing, research and development, and manufacturing. Numerous studies have shown that such teamwork is essential for a company's success.

It links the main phases of product development. A more thorough use of QFD ensures the generation of four matrices that link the four main phases of product development, specifically:

- Product planning, after customer needs are understood
- Part deployment
- Process planning
- Production planning

The QFD methodology has been around for several years. It seems to have been first used by Bridgestone Tire Company. It has been successfully used by Toyota Motor Company in product development and has helped reduce product costs, rework, and design changes after product release.[8] Toyota builds cars that better meet customer needs. Toyota has also reduced the overall time to design and market its cars. Currently, Toyota has about the shortest product development time in the automobile industry; it is about 3 years.

We recommend use of QFD methodology; details are provided in Chap. 5, "Daily Process Management."

Meeting and Exceeding Customer Needs

The various methodologies discussed so far will help ensure you have a more customer-sensitive organization. There is of course another ingredient that is necessary—creativity. We believe that the customer-oriented activities that we have recommended will form a strong foundation that provides a good basic system and will allow creativity to be channeled to the right areas.

One such area is in providing "attractive" quality or exciting features in your products and services. This is what the customer wants: exciting products and services. This concept of attractive quality can be integrated into the QFD matrix after customer needs and competitive offerings have been assessed. Refer to the discussion in Chap. 1 on definitions of quality—specifically, must be and attractive quality.

Attractive quality in products

For products, attractive quality can be easily seen. In automobiles, for example, it includes features such as antilock brakes, safety air bags, sun and moon roofs, electronically adjusted seats, quiet operation, and luxurious interiors. And, most important, all this is expected at a reasonable price. The Toyota Lexus automobile is a stunning example. Its attractive quality features are threatening luxury car leaders such as Mercedes Benz and BMW. The same concept can be applied to all products and services.

Attractive quality in services

How can we apply this concept to services? Think about a favorite restaurant or airline—their excellent and special services come to mind. Today, as we write, Singapore Airlines is planning an "office in the sky" for all its Boeing 747 aircraft. This will be a first and will include video conference facilities, desktop computers, photocopying equipment, and easy payment. This is attractive quality for the busy executive.

Here is a unique way of looking at attractive quality in products and services, with an example of how one department store provides attractive quality in its service:*

> Harvard Business School marketing professor Ted Levitt provides a handy idea, which he labels the "total product concept." Picture four concentric circles: The inner one is labeled "generic product," next comes "expected product," then "augmented product." The last, which has no boundary, is labeled "potential product" (see Fig. 2.7).
>
> Take Nordstrom, whose fairly high price specialty retail goods comprise the firm's generic trait. The expected trait includes stores that stay open normal hours and carry styles that are contemporary. Nordstrom invests heavily in its remarkable service—its augmented attribute—such as very high availability of odd sizes and colors and high pay, by industry standards, for an exceptionally large number of salespersons on the floor.
>
> The unlimited potential trait is a vast number of touches—from regular personal notes to customers from sales persons to the especially clean and colorful dressing rooms to a "no-questions asked" return policy, all of which help Nordstrom live up to its "no problem at Nordstrom" logo.

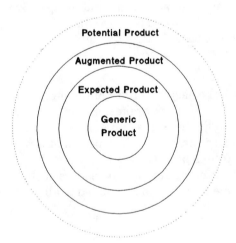

Figure 2.7 Total product concept.

---

*Excerpted from Thomas Peters' weekly syndicated column. Copyright 1987, TPG Communications. All rights reserved. (*Note:* We have provided the illustration.)

By its unconventional emphasis on the outer circles—the augmented and potential—Nordstrom has virtually redefined retailing. To quote a friend at a computer company, it's not a specialty retail store but "a user friendly, entertaining, total experience" that has something to do with the purchase of garments.

In conclusion, providing attractive quality features will allow us to meet and exceed customer needs—when designing products and services, we must consciously strive to provide these; in fact, there must be a check-off procedure that ensures that we are doing this. That will provide the key to successful products and services.

Questions and Answers on Customer Obsession

Let us review a few questions on customer obsession.

Q1. The customer complaint and management system you propose is very complex. My employees are well trained and very customer conscious; why should I start a formal system and hire more resources?

A1. An informal system has some weaknesses. You cannot be sure that every complaint is attended to because sometimes an employee is busy or pressured by other priorities. Often, the recipient of the complaint has to go to someone else for help and that process does not always work. Also, if you have a large organization, the same problem may be occurring in other parts of your organization—in this case you have a class, or companywide, issue that needs to be escalated and addressed so that the problem does not repeat across the organization. All these are reasons to invest in a formal system.

Q2. We recently conducted a customer satisfaction survey and got a very high satisfaction score, yet our market share is low. What's happening?

A2. The reasons can be complex. You may be asking the wrong questions—that is, the things you ask about may not be important to the customers; your competitors may be providing those additional needs

In measuring satisfaction, you may be talking to a captive audience. That is, customers who own your products—they may get the best satisfaction for your product and for the given set of features that your product provides; yet other competitive products may have better features and a higher value. You must measure the entire spectrum of features and activities that a customer may want. Consider capturing and collating satisfaction data in a customer satisfaction model similar to the one shown in Fig. 2.6. This will require collating data from several sources.

If you compete at a higher price or have poor marketing efforts, your market share will be lower too. To understand the root cause of low market share you need to do a detailed analysis.

Summary: Creating a Customer-Obsessed Organization

We have discussed the importance of customer obsession. Having well-trained employees with good intentions will not be sufficient to provide this obsession. We have proposed several methodologies: reactive and proactive. They include the following:

- A customer complaint and feedback system
- Customer satisfaction surveys and a customer satisfaction model
- Competitive benchmarking
- Capturing the customer's voice and needs via QFD

Some methods will make you more customer sensitive—specifically, customer surveys and the customer complaint and feedback system. Competitive benchmarking will help ensure that you are at least equal to the "world's best" companies. The use of QFD methodology will help ensure that you design the right products and reduce the frequency of failed products and wasted resources.

QFD can be integrated with the concept of designing-in attractive quality. This will propel you ahead in providing exciting products and services and will give you a competitive advantage.

All the activities listed will help create a successful, customer-obsessed organization. You need to plan for these activities and that is our next topic.

References

1. Juran, Joseph M., Frank M. Gyrna, and R. S. Bingham. *Quality Control Handbook.* New York: McGraw-Hill, 1951.
2. Camp, Robert C. *Benchmarking: The Search for Best Practices that Lead to Superior Performance.* Milwaukee: Quality Press, 1989.
3. "Competing with Tomorrow." *The Economist Magazine,* May 12, 1990.
4. Prahalad, C. K., and Gary Hamel. "Core Competencies." *Harvard Business Review,* May/June 1990, pp. 79–91.

5. The Science Policy Research Unit, University of Sussex. *Success and Failures in Industrial Innovation.* Edinburgh: Bishop and Sons, 1972.
6. Doyle, John. Quoted from a speech for Hewlett-Packard Quality Managers. Santa Clara, California, May 1990.
7. Timmins, E. Scott, and Jack Solomon. "Quality in R&D: How to Involve the Customer and Like It." *1900 ASQC Quality Congress—San Francisco.*
8. Sullivan, L. P. "Quality Function Deployment." *Quality Progress,* June 1986.

Chapter

3

The Planning Process

For I dipped into the future, as far as human eye could see, saw the vision of the world, and all the wonders that would be.
Locksley Hall
ALFRED, LORD TENNYSON

Overview

Preparing for the future is crucial. In today's highly competitive and changing marketplace the margin for error is decreasing; hence planning for the future is necessary for survival and success. Planning and executing the plan are key activities in a company and are done by its managers. If done well, this process will give the desired results.

Formal planning provides many benefits, including systematic thinking, better coordination, sharper objectives, improved performance standards, and management involvement. All of these result in a planned approach to tackling the marketplace that can end in higher sales and profits.

The following incident indicates what can happen when planning is not formalized in a company.

> I visited a large sales operation to conduct a quality review. On the second day I met with a district sales manager and his staff. He was extremely intelligent, young, and enthusiastic. During the current fiscal year he had lost his major account—they had been acquired by another company and their purchases from his company declined very rapidly. What to do? He decided to prepare a plan and purchased a copy of a book entitled *One Page Management*. He read it and started training his staff in it. In addition, he started a process to manage his sales funnel—the performance measures and goals to manage this funnel were entered into his one-page management plan.

He had been experimenting with this plan for several months and was still fine tuning the performance measures. I asked him why he did this. He replied that he knew he was in trouble when he lost his major account and would not meet his quota for that year; he decided he had to plan—do something different—in order to get his sales up and his team motivated. The book helped him get organized and he felt he was now well on the way to meeting quota.

I remarked that a planning process, better than what he had, existed in the company; also I personally had written a book that addressed how the sales funnel was managed in one of the country operations in his company. To my astonishment he had never seen or heard any of this, which is why he had been forced to develop his own planning process.

The moral is simple: He managed by experience; the lucrative major account made him complacent. With the loss of the account he had to change his ways—he went into a panic mode and then developed a plan. He had wasted several months of valuable and irreplaceable time. Most of this could have been avoided if the company had standardized the planning process, trained him in it, and taught him how to manage his sales funnel. If you don't have a good planning process, your employees will design their own—and it may not be the best.

Essentials of a Planning Process

A good planning process needs to exist throughout an organization. It should be a standard process, and everyone should be trained in it. Once in place it provides a common planning language with common formats. As managers and employees transfer or move within the company, they need only worry about the planning content, the process having been standardized.

A good planning process will consist of a long-range plan and an annual plan. The long-range plan includes an analysis of the current situation (reviews of strengths and weaknesses, competitors, customers, opportunities) and then defines broad objectives and strategies that must be pursued. These objectives, which are fact and data based, are aggressive and crucial for success and are achievable given the right resources.

The annual plan is a detailed version of the first year of the long-range plan. It is a statement of intent. It includes the breakthrough objectives and implementation plans for these objectives. More specifically, it includes the how, who, and when statements of each objective. For this portion, we recommend the use of Policy Management, better known as "Hoshin Kanri."[1]

The combination of the long-range and short-term plans forms a powerful and planned approach to succeeding in the marketplace. Our discussion will focus in detail on the annual plan, but a short discussion on the long-range plan will be appropriate.

Long-Range Plan

An organization must understand its purpose for existence, the marketplace, and the competition and then must determine its long-term direction. One company in particular, Matsushita of Japan (owner of the Panasonic, Technics, and National brands), is reported to have a 250-year plan. This was developed in 1932, and by all accounts they are progressing to plan; the plan is reputed to include company growth, products, markets, and countries covered. In our context, we are proposing a 3- to 5-year plan. The long-range plan should cover the following steps and areas:

1. *Company's or organization's purpose and vision.* Before any planning starts, the organization's purpose and vision should be prepared and clearly understood. The purpose and vision will describe the fundamental set of reasons for the organization's existence; it should be inspiring, give a clear sense of direction, and provide a basis for decision making. It should also indicate where you will be in the future and how you expect to provide a competitive advantage. And it should be simple. Let us look at some examples. Here is part of the Honda Motor Company's vision: "...supplying products of the highest quality yet at a reasonable price for worldwide customer service." And here is part of Toyota Motor Company's vision: "Be number one in quality and offer good products to respond to requests of society and trust of customers."

Both of the visions are simple and include a quality or customer satisfaction statement. We recommend appending specific goals to the vision and purpose. These goals can be valid for, say, 5 years. After they are achieved, they can be reset with new goals. So, for example, both the purpose and visions could have a quality goal which states: "be among the top three in customer satisfaction." More on this in Chap. 7.

2. *Customer needs, issues, and channels of distribution.* Here we conduct a rigorous analysis, based on data, of customer needs and trends. For example, what problems are customers trying to solve? What are their dreams? This allows us to discover new market segments and to build a data-based model of the market. Our focus will be within our purpose and vision—often, this will change as we enter new markets. When marketplace needs are understood, they can be grouped by market segments and areas of opportunities. In addition, the market segments must be mapped against current and future channels of distribution. To reiterate, we must understand the following:

- Customer needs
- Market segments
- Channels of distribution

3. *Competitive situation.* Here the focus is on a competitive analysis. We need to understand the competition, their strategies, how they differentiate their products and services, and their financial strength. In addition we need to know their strengths as perceived by their customers, how their strategies have evolved over the years, and their current strategies.

4. *Products and services.* Based on customer needs and target market segments, we determine our proposed products and services. This includes continuing with current offerings and planning for new offerings. Key technologies, or core competences, must also be identified and planned for.

5. *Development of partners and purchase plan.* We define a plan for developing products and services that we need to purchase from third parties and partners. In addition to products, this may include services, such as after-sales support and product distribution.

6. *Financial analysis.* Here we do an analysis of revenues, cost of goods, expenses, investments (including R&D), overhead costs, and profits. We must also determine the return on investment for the proposed products and services. Five-year projections for these items should be available.

7. *Potential problem analysis.* It is important that a risk analysis be conducted and appropriate contingency plans proposed. Areas of risk and the possible competitive response should be reviewed.

8. *Five-year plan.* We are now ready to list the 5-year plan according to the major functions in the organization. For a manufacturing company this would include product plans and objectives for marketing, design, and manufacturing. Also included would be plans for quality and human resource development. This step should include introduction of new strategic products and development of core competences.

Here is an example of breakthrough, long-term, objectives from Northern Telecom.[2] Faced with increasing competition, they found that customers wanted higher quality and more sophisticated products, at a low cost. In 1985, they defined three objectives for the next 5 years. These were:

- Increase customer satisfaction by at least 20 percent as measured in the annual survey
- Reduce manufacturing overhead as a percentage of sales by one-half

- Cut inventory days by 50 percent.

Their fundamental thrust was to double throughput velocity without raising expense or inventory levels while substantially increasing customer satisfaction. The data shows that Northern Telecom is progressing well to plan.[2]

9. *Annual plans.* Finally, we prepare the annual plan. The Hoshin planning format is recommended. The Hoshin plan will include items from the 5-year plan as well as other items, which are discussed in the next section.

The output of this nine-step planning process is a long-range (3 to 5 years) plan and the annual Hoshin plan. This entire process must be repeated every year because markets and competitors change or the economic environment may vary. Typically there will be fine tuning of the plans throughout the year with a major revision each year. In some years, if there is little change in the market or competition, the 5-year plan may require only minor adjustments. The annual Hoshin plan, however, is done anew each year.

Annual Plan

Every organization must have an annual plan. A good annual plan will move the organization to a better position in the marketplace each year. And yet, we have found a dearth of literature on a detailed annual planning process—details such as process flows, formats, planning guidelines, and training material. Even business schools do not put much emphasis on an annual planning process. Therefore we will discuss this topic in detail.

We recommend the use of the Hoshin Kanri and Nichijo Kanri methodology. This is a very successful planning process used by almost every Japanese company that has won the Deming Prize. It originated in the 1960s at the Bridgestone Tire Company in Japan. At that time, weaknesses in their planning process were becoming apparent.

The Hoshin Kanri philosophy originates with ancient military traditions and efficiency. Many large Japanese companies and several American companies that adopted *management by objectives* (MBO) have given up that process and switched to Hoshin Kanri type planning. Details of Hoshin Kanri and comparisons to MBO will be given in the following discussions.

The terms *Hoshin Kanri* and *Nichijo Kanri* are loosely translated from the Japanese as follows:

Hoshin Kanri. Hoshin means objectives or directions, while Kanri means control or management. So in essence Hoshin Kanri means policy management or management of objectives (contrast this to management by objectives in the Western corporate environment). Henceforth let us just call it the Hoshin plan.

Nichijo Kanri. Nichijo means daily. Hence Nichijo Kanri means *daily management.* In the Western corporate environment a clearer definition could be business fundamentals. We will use the terms interchangeably but prefer the term *Daily Management*; the reason will become apparent later.

A brief description of the Hoshin planning process is given below. In the succeeding pages the various formats to be used are described and suggestions to develop a good plan will be given with examples.

Hoshin planning process

The process begins with a review of the following items (refer to the planning flow in Fig. 3.1):

Company purpose and vision. The company's purpose and vision and the long-term objectives, which are regularly updated, are reviewed.

Specific customer inputs. Specific customer inputs, issues, and needs that have been collected during the year are reviewed. Customer inputs are important—these can come from regular customer surveys, meetings, visits, letters, and customer data gathered during preparation of the long-term plan. The need for closeness to customers can never be overemphasized.

Current economic situation. The current economic situation and its impact is reviewed.

Review of the previous year's plan. Successes and failures in the previous year's plan are reviewed at all levels of the organization. The lessons learned will influence the upcoming year's plan. This is an important step and will be discussed later in the section entitled "Hoshin and Business Fundamentals Plan Reviews."

All of the above items form a list of key issues from which the Hoshin plan will be prepared. Typically, the entity general manager (which can be the CEO/president or a product or service division manager) will prepare the annual Hoshin plan. This plan will provide broad objectives, strategies, and performance measures. The next-level managers—marketing, finance, research and development, operations or manufacturing, and quality—will then prepare a more specific plan, after discussions with their department managers. Refer again to the Hoshin planning flow chart in Fig. 3.1. The department managers will then prepare more detailed and specific plans. In addition the department managers will prepare an implementation plan.

The Planning Process 55

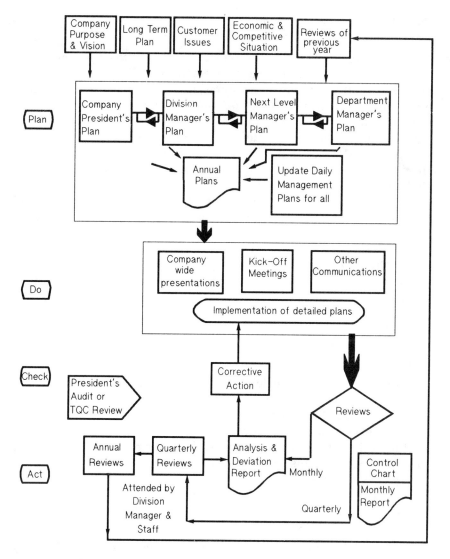

This chart shows the sequence of activity for generating a Hoshin Plan and a Daily Management/Business Fundamentals Plan. Note that the double arrow symbol (▶◀) denotes a management discussion mode, which is crucial in ensuring good deployment of plans.

Figure 3.1 Hoshin planning flow chart.

Daily Management, or Business Fundamental, plan

So far the Hoshin plan covers the objectives that flow down from upper-level management. These indicate what the entity must achieve in the year. But what about day to day items, items that must be maintained and monitored? These include cost controls, employee morale, selling or manufacturing a product, maintaining or slightly improving a product or process, or some other annual repetitive task.

The Daily Management, or Business Fundamentals, plan addresses this issue and focuses on "keeping the house in order." There is very little higher-management-level involvement in preparing this plan, but all levels of management should have one. The final plan should have both Hoshin and Daily Management plans.

Launching the plan

Next, the plan is launched. Refer to Fig. 3.1. Often a facilitywide meeting is held to present the plans, but at a minimum the plans should be presented to all management and supervisory staff.

Hoshin plan reviews

Hoshin plan reviews are held regularly—preferably every quarter or 3 months. If everything is on track, it is business as usual. If not, plans may have to be changed or more resources acquired. Finally, the results and experiences of the current year are summarized in an annual Hoshin review. The annual review is done during the last quarter of the planning year; this sets the stage for starting the next year's annual planning cycle.

Hoshin plan: Ensuring success

The strength of the Hoshin and Daily Management planning process is a systematic and tightly coupled process. This process ensures three things: First, that the plan can be achieved because the next level has committed to it. Second, that there is a hierarchy of objectives and strategies. Third, and most important, that the plan is reviewed regularly and corrections made to get the entity or ship back on course.

Illustrating the Difference between MBO and Hoshin Planning

We have discussed the difference between MBO and Hoshin planning. In Fig. 3.2 we illustrate the difference between alignment of objectives in MBO and Hoshin planning. As the figure shows, the use of Hoshin Kanri planning ensures better linkage and alignment of objectives at the various levels of management in the company. This is due to the Hoshin planning methodology, which ensures very tight cascading of objectives from one management level to the next. In fact the original Kanji (Japanese) characters of Hoshin Kanri mean "shiny needle or compass." Today, we know of several managers who refer to Hoshin Kanri planning as compass management—that is, everybody managing and moving in the same direction.

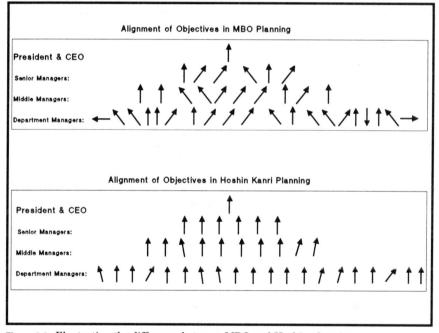

Figure 3.2 Illustrating the difference between MBO and Hoshin planning.

Relation between Hoshin plan and Daily Management plan

At this juncture, it will be useful to discuss the difference and relationship between Hoshin planning and Daily Management planning.

Imagine a spaceship taking passengers on a journey from earth to the moon. Its Hoshin plan addresses how it gets from earth to the moon—planning and managing the journey through space and watching out for meteors and aliens. At the same time its Daily Management plan addresses the day to day task of feeding and entertaining the passengers and keeping the spaceship well maintained and running smoothly. Obviously you need both types of plans, and the separation into Hoshin and Daily Management is very useful for directing management focus, allocating appropriate resources, and setting priorities.

MBO or Hoshin planning?

You may ask the question, "Why adopt Hoshin planning? After all it is very similar to MBO and we have been successful with MBO." While MBO has many strengths, it has also many weaknesses. For example, there is a weak linkage between strategy and implementation; there is no detailed planning process; there is an insufficient consensus approach; a hierarchy of objectives, although apparent in theory, may not exist in reality; finally and most important there is no framework for a formalized review procedure to monitor and ensure success.

Hoshin planning on the other hand has all the strengths of MBO and more to boot but none of its weaknesses. The strength of Hoshin planning is that it is a systematic and tightly coupled process. It does, however, require much more effort and consensus than MBO; but it helps provide a focus, a single-minded approach by the entire management team. The entire process is designed to ensure success. In the final analysis, Hoshin planning can be considered a more mature MBO process.

Formats and Guidelines

Next, let us review some recommended formats and guidelines for the Hoshin planning process and Daily Management.

Annual Hoshin plan

The annual Hoshin plan summarizes the breakthrough objectives for the entity or organization. These could be objectives that let you leapfrog your competition. Accomplishing these objectives normally requires more than the ordinary sustaining effort and is likely to involve multi-department collaboration. The plan generally comprises four elements:

1. Objective
2. Target or goal
3. Strategy
4. Performance measure

A recommended format to capture these elements is shown in Fig. 3.3. The following is an explanation of each element. Completed examples with comments are shown in Figs. 3.4 and 3.5.

Objective. This is a purpose to be achieved. Usually an aggressive or breakthrough statement.

Target/goal. This is a broad indicator measuring accomplishment of an objective. It must be established for every objective and be quantifiable.

Strategy. This describes the procedure and method by which the targeted goal is to be accomplished.

Performance measure. This is used to determine the progress or completion of a strategy. It consists of a statement and a number; the number indicates the target to be achieved.

Let us now examine some of these elements in detail, namely: objective, strategy, and performance measure. We will start with the most crucial: the objective.

Objectives. Before objectives are prepared, you should develop an issue list. Refer to the Hoshin planning flow chart in Fig. 3.1. This starts with a review of the following:

- Company vision and the long-term, say, 5-year objectives. These are regularly updated.
- Specific customer inputs and issues collected during the year.
- The current economic situation.
- Successes and failures in the previous year's plan at all levels in the organization.

ANNUAL HOSHIN PLAN			
Objective	Target/Goal	Strategy (and owner)	Performance Measures

Figure 3.3 Annual Hoshin plan.

ANNUAL HOSHIN PLAN

	Prepared by: General Manager	Division/Department Apex		Fiscal Year 1989	Page 1 of 2

OBJECTIVE	NO.	STRATEGY (OWNER)	PERFORMANCE MEASURE
1. Increase revenue while improving Customer Satisfaction	1.1	Grow our average account's annual dollar volume and individual order size for our direct sales in computer systems (Marketing & Sales Managers)	* Increase business by 20% for top 30 accounts * Penetrate 10 new Fortune 100 accounts in our area
	1.2	Improve capability and efficiency of Quality Assurance Department in the Computer Systems environment. The focus should be in eliminating current customer concerns and issues in computer systems (Quality Manager)	* Solution ratio of major Customer complaints at 90% * Customer Satisfaction level improved by 10%

TARGET/GOAL

* Orders at $ 1 Billion (a 25% increase)
* Orders at target every quarter

Here are some comments on the development and deployment of these hoshin plans:

* The Division/Plant Manager develops his/her annual plan as shown on left. The appropriate next level manager then picks whichever strategy is appropriate to him/her and makes that his/her objective.
* Note that the General Manager's First Objective attempts to capture both Quality and Delivery from the OCDE categories. This is reflected in his / her strategies, where both points are further developed. A discussion on OCDE is provided later in the text.
* In this example, the Quality Manager has been assigned Strategy No. 1.2, which becomes his / her Objective and it's Performance Measure becomes the Goal. In a similar fashion all the General Manager's Strategies will be deployed to the next level.
* Another example on Hoshin plan development is given in the next figure.

ANNUAL HOSHIN PLAN

	Prepared by: Quality Manager	Division/Department Quality/APEX Division		Fiscal Year 1989	Page 1 of 2

OBJECTIVE	NO.	STRATEGY (OWNER)	PERFORMANCE MEASURE
1.2 Improve Capability and structure of Quality Assurance throughout the computer system business.	1.2.1	Improve solution ratio of major customer complaints by setting up a Quality council	* Analyze and issue corrective action requests for all major complaints
	1.22	Establish formats and system for capturing customer quality problems for all computer products and be able to relay them to the production line quickly	* Deviation from implementation plan * Data arrival at production line within 36 hours of customer complaint
	1.2.3	Reduce the recurrence of all complaints of customers	* Reduce recurrence rate by 50%
	1.24	All Dept. Managers to select one project for improvement	* Project agreed to by Quality Manager * Project completed in 9 months
	1.25	Determine Training Needs	* Determine by March, 1989 * Provide Training by 2 Half 1989

TARGET/GOAL

Goal #1: Obtain a solution ratio for major customer complaints of 90%

Goal #2: Customer Satisfaction level improved by 10% in survey.

Figure 3.4 Preparation and deployment of a product division manager's plan.

ANNUAL HOSHIN PLAN

Prepared by: General Manager	Division/Department: APEX	Fiscal Year 1989	Page 2 of 2

Situation Analysis

The APEX Division profits have been low. Mostly this has been due to an old product line, high cost structure and delayed introduction of new products. The new product line will increase sales and profits. However analysis shows we must reduce current manufacturing and administrative costs by 30 & 15% respectively. Field Failures must also be reduced below competitive levels. In all this will boost profits from 3 to 6%.

OBJECTIVE	NO.	STRATEGY (OWNER)	PERFORMANCE MEASURE
2.0 Increase Profits	2.1	Introduce New Products per schedule (R & D and Marketing)	* New products to provide 30% of Total Sales in 1989
	2.2	Reduce Costs in Manufacturing and Administration (Manufacturing Operations and Administration Department)	* Reduction in Manufacturing costs by 22% * Reduction in Administrative costs of 15%
	2.3	Reduce Product Field Failure rates and thereby minimize warranty costs (Manufacturing Operations)	* Reduce Failures from 4% to 1.5%
	2.4	Etc.	

TARGET/GOAL

Increase from 3% to 6% (minimum acceptable is 5 %)

Here are some comments on the preparation and deployment of these Hoshin plans:
* The Division General Manager has selected profits as his no. 2 objective. This is the corollary of C from the OCDE categories. In addition the strategies for the objective include Q, C, and D.
* Note that in this example all goals have limits. That is, a minimum acceptable level of performance (refer to the discussion on limits).
* The Operation Manager has selected the strategies that are deployed to him/her; but has selected more aggressive goals, with limits equal to the General Manager's performance measures.
* A Hoshin deployment matrix, which is shown in the next figure could be used to determine ownership of the General Managers strategies and to agree on who does what.
* Note also that there is a numbering system that provides traceability of Objectives and strategies. For example, the General Manager's Objective is no 2.0; the strategies are no.2.1, 2.2, etc; the Operation Manager's objectives take on the number of the General Manager's strategies - 2.2 and 2.3; and the strategies become 2.2.1, 2.2.2, etc.

ANNUAL HOSHIN PLAN

Prepared by: Operation/Manufacturing Manager	Division/Department: Apex, Manufacturing	Fiscal Year 1989	Page 1 of 2

OBJECTIVE	NO.	STRATEGY (OWNER)	PERFORMANCE MEASURE
2.2 Reduce Manufacturing Costs	2.2.1	Review and reduce procurement costs (Procurement/Materials Manager)	*Cost reduction for Product A, B & C of 15%
	2.2.2	Redesign and Simplify Product A and B. Their products have an expected life of 3 more years (Engineering Manager)	*Redesign cost should be 15% less for A and B
2.3 Reduce External Product Failure Rate	2.3.1	Improve Yield of current products (Production Manager)	*Yield of A: 85 → 92% *Yield of B: 95 → 98% *Yield of C: 98 → 99%
	2.3.2	Improve Field failures based on pareto of defects (Engineering & Production Manager to form a cross-functional team)	*Reduce top 5 failures by 50%
	2.3.3	Etc.	

TARGET/GOAL

For 2.2 Cost Reduction of 30% (minimum acceptable will be 22%)
For 2.3 Fail Rate reduced from 4% to 1% (minimum acceptable will be 1.5%)

Figure 3.5 Preparation and deployment of a product division manager's plan.

The last point listed is extremely important—successes and failures of the previous year are reviewed and analyzed before a new plan is prepared. In the boxed example (p. 63), we show an issue list generated from the previous year.

The following objectives will determine the direction in which your business is heading. As a guide, objectives should cover four important categories, abbreviated as QCDE:

Quality (Q). Includes customer satisfaction issues and product/process quality.

Cost (C). Includes all costs, such as administrative expenses, manufacturing costs, and productivity issues. The corollary to this is profits, which is, of course, of paramount importance.

Delivery (D). Includes new product design introductions, R&D and manufacturing product commitments, and delivery of products to the customer.

Education (E). Human resource training and education, and organizational issues.

The concept of QCDE is meant to ensure that nothing important is overlooked. *We do not suggest that you have four objectives—we only suggest you review these four categories. In fact we recommend that senior managers do not have more than two objectives—with several supporting strategies.* In actual practice, some of the objectives can have strategies that address more than one of these categories. For example, an objective to increase profits can address costs, delivery, and quality. Refer to the example in Fig. 3.5. Some companies that use Hoshin planning actually have prompts on their planning forms to ensure that managers consider these four categories.

Before we go on there are several points to be made on the concept of QCDE.

1. In TQC philosophy the QCDE categories are considered essential for business success; progress in each category will help keep a company robust, healthy, and competitive. Complacency in any category may cause problems. After generation of the key issues, which was discussed earlier, the issues can be categorized in the QCDE categories.

2. Ideally, the number of objectives should be limited to no more than two or three. This may be difficult at first but can be achieved if a manager focuses only on "breakthrough" objectives—that is, objectives that are essential for success. Experience has shown that an individual manager can only focus on two or three major objectives

> **An Example of an Issue List from the Previous Year**
>
> In preparing a Hoshin plan, you need to review the previous year's performance at the end of that year. What can you expect from an end of year review? It would be a list of problems or lessons learned from a job well done. This list of issues can influence the following year's Hoshin plan. Here are some examples:
>
> **For a sales and marketing operation**
>
> - Lost 35 percent of big deals when competing for a sale of product A1234. There is a need to understand what happened—is it a product or marketing problem?
> - Won 95 percent of big deals when competing for a sale of product B224. Why did we do well? Can we transfer the lessons learned to other new products and marketing?
> - High inventory of demonstration units—now obsolete. What happened?
> - Sales representatives are inexperienced because of extensive hiring during high growth. Need training next year.
> - Management quality team improvement projects were ignored. What happened?
> - High inventory of 2125 product. It seems market forecasts may be the cause—need to investigate.
>
> **For a design and manufacturing environment**
>
> - Product C234 process yield fell because of the purchase of inferior semiconductors. Created problems with customers. What happened?
> - Management quality teams were very successful. Need to expand the concept to other areas, including sales.
> - Product B224 was very successful, whereas product A1234 was not. Need to understand success and failure factors.

at a large organization. To try to do more may cause you to overstretch or loose focus.

3. In any year, not all the QCDE categories need to be addressed. For example, if in the previous years you set up a comprehensive quality education and training program, you would not worry about it in the current year because it is now a Daily Management item. The personnel or human resources department is managing this on a day to day basis.

4. Often, the QCDE categories will merge. For example, a profit objective could include strategies for quality, costs, and delivery—as shown in Fig. 3.5.

5. In setting the objectives a hierarchy is important. So for the Q category, the general manager of the organization could set a top-level customer satisfaction objective with a broad overall customer satisfaction goal—say the result of an industrywide survey. But at the lower level a product or service manager would have a specific ob-

jective that supports the top-level objective and goal—say, an objective to improve a product or service that customers are unhappy about. A similar analogy will apply for the other QCDE categories.

6. The C category is often overlooked in many organizations. This often becomes a concern only just before or after a company takeover, acquisition, management change, or a sudden downturn in profits. When we mention costs, we do not mean just specific product or service costs. We mean *all* costs. And this includes overhead and other costs which have a tendency to creep up, especially during good times. Ignoring costs—that is, not managing them on a year to year and a daily basis—can result in the *self-feeding cycle of competitive decay*.[3] One of the ways of reducing costs is to improve quality; this is addressed more specifically later. But other areas of equal importance are administrative and overhead costs, which must be managed on a routine basis, year in and year out. See the box below for a detailed discussion on managing costs.

Chief executive's objectives

We have given some guidelines for selecting and grouping objectives—basically using the concept of QCDE. If you are managing a complex organization—say, several divisions or several countries—then it may get more difficult to find a common denominator. In addition, you must be strategic and not have too many objectives.

What is too many? Typically, if the chief executive has more than two objectives, it is too many. Experience has shown that one or two breakthrough objectives are about the maximum that an organization can manage.

Managing Costs

A recent article in the *Harvard Business Review*, "Vital Truths About Managing Your Costs" by Charles Ames and James Hlavecek,[3] has an interesting discussion on managing costs. The authors state that there are several truisms that apply universally in the business world. Two of them are:

1. Over the long term, it is absolutely essential to be a lower-cost supplier.
2. To stay competitive, inflation-adjusted costs of producing and supplying any product or service must continuously trend downward.

In particular, product costs include overhead or other costs like designing, selling, delivering, and service and must be managed. Especially dangerous is the fact that these tend to overaccumulate in good times when there is no pressure for tight performance and good sense. In particular, they stress that most costs are manageable, while only a few are truly fixed. Ames and Hlavecek go on to discuss the self feeding cycle of competitive decay, and we show a copy of their chart in Fig. 3.6. This viciously deteriorating cycle can work itself out into worsening conditions and must be prevented.

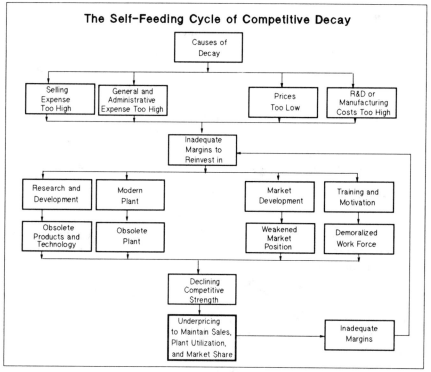

Figure 3.6 The Self-Feeding Cycle of Competitive Decay. Copyright © 1990 by the President and Fellows of Harvard College; all rights reserved.

Another good reason for the chief executive to have few objectives is that this will allow lower-level managers to add objectives that are specific to their situation. Otherwise, you could have a situation where the chief executive has, say, four objectives and the general manager adds a few more. Very soon the operation will be awash in plans and objectives, resulting in a loss of focus.

Our recommendation for the chief executive is to follow these steps before defining his or her objectives:

1. *Look for common issues or problems.* For example, in the area of quality and customer satisfaction, or profits
2. *Review specific needs of a weak area in the organization.* For example, a weak product line, division, or country
3. *Review future needs in your organization.* For example, startups in new countries, new markets to penetrate, new products to be introduced, or new technologies to be acquired

With this guideline, the chief executive will be able to have one or two breakthrough, specific, and focused objectives for the organization.

Strategies and performance measures. We have talked about objectives; now let's discuss strategies and performance measures.

Strategy. A strategy describes the procedure and method by which the targeted goal is to be accomplished. An average of three to five strategies is recommended for each objective. However, the actual number can vary depending on the complexity of the objective. Too many strategies could result in loss of focus and hence loss of control.

Is there a technique for generating strategies? Yes. We recommend two methods that have been used successfully by many managers:

Generating strategies: method 1
1. List each of the objectives with its goals.
2. Generate a list of strategies (by brainstorming or other means) by which the objective can be met. Many of them may be apparent, but you may need others.
3. Evaluate the choices by ranking each according to its contribution toward fulfilling the objective, its cost, its feasibility, and other important limiting factors. If necessary you can weigh each strategy as follows: contribution × cost × feasibility; but this is seldom necessary.
4. Establish ranking order of the choices.
5. Select the choices based on ranking.
6. Discuss the selected choices with managers to obtain endorsement. The discussion nodes shown in Fig. 3.1 are meant for this, and this is of crucial importance.

Generating strategies: method 2. A second method of determining appropriate strategies is a more scientific and data-driven method. As we have mentioned previously, the Hoshin planning process is a large PDCA cycle—and preparing a Hoshin plan is the P stage of this cycle. Basically, this is what you do:

1. Once the Hoshin plan objective is determined, prepare a cause and effect diagram—where the effect is the selected objective. Many of the causes can then be be brainstormed.
2. Determine the most likely causes that affect the objective and verify them with data.
3. Convert the verified causes into strategies for the selected objective.
4. Ensure that the strategies, after implementation, add up to meeting the objective and goal.

Let us review an example of how this is done. Suppose one of the selected objectives is "to increase profitability." Then a cause and effect diagram for "why are profits low?" can be constructed. An example is shown in Fig. 3.7. Next, the most likely causes are selected, discussed, and verified with data. These have been marked with a check in the cause and effect diagram. The three most likely, and verified, causes could be:

- Lack of new products
- Product failure rate not optimum
- High manufacturing and administration costs

Appropriate strategies can now be prepared to reduce or eliminate these causes. In fact, we have done this in the Hoshin plan example shown in Fig. 3.5. In the example, the strategies add up to the goal. This is not immediately obvious; but the situation statement at the top of the plan refers to a separate financial analysis. In the analysis, the specific performance measures were selected to ensure that the goal would be met if the performance measures were met.

This second method is very useful in formulating robust and effective strategies. Certainly, it is better than the standard strategy, these days, for improving profits—that of reducing people. This only gives short-term results and, worse still, may not get to the root cause of the problem.

For more details on cause and effect diagrams, refer to the discussion in App. 2 on the seven tools. For more details on PDCA and verification of causes, refer to the P stage of the detailed discussion on the PDCA cycle in Chap. 4, "The Improvement Cycle."

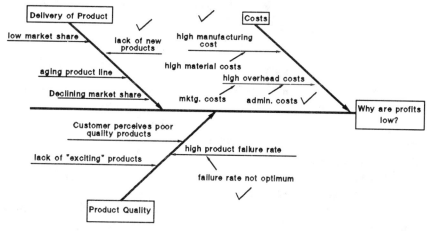

Figure 3.7 Cause and effect diagram for low profits.

Performance measures and types of performance measures. A performance measure is used to determine the progress or completion of a strategy. It consists of a statement and a number; the number indicates the target to be achieved. There are two types of performance measures, which are listed below:

Result-oriented or end-of-process performance measure. A way to measure the outcome or desired result of the strategy; for example, an action plan, increased sales or higher product line yield

Process-oriented or in-process performance measure. A way to measure progress of that strategy; for example, phased results, interim action steps, or targets in various steps of the process

Where possible try to have one of each, but this is not always possible. In that case, do the following: At senior management level a result-oriented measure will be sufficient; at lower levels a result-oriented measure will be supplemented with an implementation plan, which will provide details on the various phases of the strategy.

Situation analysis. In preparing objectives and strategies a detailed review of the current situation is important. In the Hoshin flow chart, Fig. 3.1, we show the various items that are referred to prior to preparation of the plan. In the section "The Hoshin Planning Process" we have given some more details. Such a detailed analysis is crucial in selecting the few breakthrough objectives you wish to select. In the analysis, you should also be able to determine the appropriate goals and performance measures to select. Such an analysis must be summarized and documented. In Fig. 3.5, we show a slightly different Hoshin plan form (different from Fig. 3.4). Here we have included a brief situation analysis statement. A good analysis should show:

- Why the objective was selected.
- Why the accompanying strategies were selected.
- How the goals and performance measures were determined.

And if necessary it should refer the reader to a separate financial or root cause analysis.

Deployment and cascading of objectives

As Hoshin plans cascade down an organization, this is what will happen: The senior manager's plan will be deployed to his or her managers. These next level managers will select the strategies that are appropriate to them—these become their objectives and the performance measures become their targets or goals; each of these objectives will generate a number of new strategies. The concept is illustrated in Fig. 3.8. Further discussions are given below and in Figs. 3.4 and 3.5.

The Planning Process   69

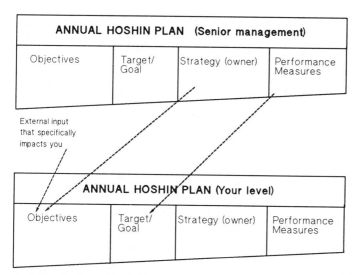

Figure 3.8   Deployment of objectives.

Deploying objectives in a large organization. Deploying objectives in a large organization is easy and effective with the Hoshin planning process. Certainly it is much easier than with any other planning process that we know of. We illustrate the concepts of deployment and cascading in Fig. 3.9. In the figure, we show how each strategy becomes the objective for the next lower level. The cascading process continues downward until the last set of strategies is reached—these end up with implementation plans.

DEPLOYING OBJECTIVES IN A LARGE ORGANIZATION
(With examples of cross-functional team and non-deployed strategy)

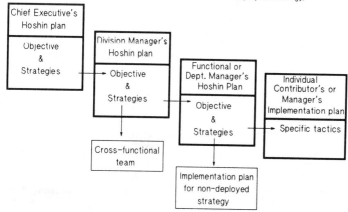

Figure 3.9   Deploying objectives in a large organization.

But not all strategies cascade down the organization. Some strategies may not be deployed and instead are picked up at a high level—we show this in Fig. 3.12, where the quality manager implements his or her own strategy. Other strategies will involve many departments and are assigned to a cross-functional team. Cross-functional teams will be needed for strategies that require participation by many functions. For example a strategy to introduce a new product will require the involvement of R&D, marketing, manufacturing, quality, and other functions. We illustrate all this in Fig. 3.9. There is more on cross-functional teams later in this section.

For this process to be effective, the chief executive's plan has to have only one or two breakthrough objectives—because at each lower level the objectives will multiply. If done well, the result will be a very tightly knit plan, across the company, with everyone moving in the same direction.

Impact of cascading and how to prevent objectives from repeating. As Hoshin plans cascade down the organization, a number of things—good and bad—will happen. The number of objectives will quickly multiply, but if they are prepared well, there will be strong linkage and tremendous synergy, resulting in a sharply focused organization. This gives a strong argument for starting with fewer objectives at the top. Unfortunately, if top-level objectives are too tactical, objectives will start to repeat. Or there can be other problems. Here are some guidelines for avoiding repeating objectives.

1. *Problem.* Objectives repeating because they are too tactical. For example, if we have a five-star general or chief executive starting with a very tactical plan such as "Take the hill" or "Hire five sales executives or ten production operators," the cascading process will not work; the result will be objectives that repeat at many levels.

    Solution. Because of the nature of the cascading process, the top-level objective (and its supporting strategies) must be broad and as high level as possible. If this is done, the lower-level objectives (and supporting strategies) can be more specific. For example:

    - At the highest level you could have "Win the war," "Increase customer satisfaction," "Increase market share," or "Increase profits."
    - At the next level you could have "Destroy Xanadu," "Improve product quality," "Add a new sales force," or "Decrease costs."

2. *Problem.* Objectives (and supporting strategies) repeat at different levels if there are too many layers of management in an organization.

Solution. This is more difficult to resolve but nevertheless needs to be resolved via a reduction in management layers.
3. *Best method.* The best way to get good cascading of plans, which are compact and concise, is to stipulate only a few layers of Hoshin plans in an organization—say, two or three. This is then supported by implementation plans.

Use of a Hoshin plan deployment matrix and cross-functional teams. Look again at Fig. 3.5. We have shown the annual Hoshin plan of a general manager (of a division or plant). Her objective is to increase profits, and three specific strategies have been listed: To introduce new products, to decrease product failure rate, and to reduce costs. The question that comes to mind is: To whom are these strategies deployed? One method is for the general manager to list the owner's name against each strategy—the owners are agreed upon during the discussion nodes shown in Fig. 3.1.

An elegant alternative is to use a Hoshin plan deployment matrix. This can be a very useful tool to ensure effective deployment of higher-level objectives. An example and a discussion of such a matrix is given in Fig. 3.10. Included in the matrix is a list of cross-functional teams necessary for success of specific strategies. Also listed in the matrix are

Hoshin Deployment Matrix							Division: APEX	Page 2 of 2
Objective No. 2 Increase Profit	Functions and Departments						Related Activity	
	R & D	Marketing	Finance & Administration	Operation/Manufacturing				
				Procurement	Engineering	Production	Cross-functional Team & leader	Quality or Project Team per department
Strategies:								
2.1 Introduce new products per schedule	◎	◎	○	○	○	○	Yes led by Marketing Manager	None
2.2 Reduce costs	△	○	○	◎	◎	◎	None	5 Teams. One each in all except R&D
2.3 Reduce external product failure	△	△	△	○	◎	◎	None	4 Teams in Manufacturing
Key: ◎ : High relationship, ○ : Medium relationship, △ : No relationship								

Figure 3.10 A Hoshin deployment matrix. Such a matrix can be used to deploy a product division general manager's objective and strategies. The matrix can be used to determine who (function or department) picks up the various strategies that were shown in Fig. 3.5. This way there will be little confusion; and strategies and performance measures can be better formulated. Later, in Chap. 4, we will show the deployment of strategies 2.2 and 2.3 to the operations/manufacturing manager's department. Note that the matrix also lists which strategies require cross-functional teams, as well as which strategies will be managed by department-level quality teams.

quality teams that will manage specific strategies. These teams can present their progress during the quarterly Hoshin plan reviews.

It is crucial that the management team determine which strategies require cross-functional teams, which consist of representatives from different functions, such as R&D, marketing, manufacturing, and so on. Strategies that require multifunctional involvement will typically require such a team. In the example in Fig. 3.10, the strategy of introducing new products requires an entitywide team effort.

The features of a cross-functional team include the following:

- It is led by a senior manager and will have representatives from all the functions that can affect the strategy.
- It agrees on allocation of responsibilities, tactics, and specific goals for individuals and departments.
- It meets regularly to measure progress, reset priorities, and request more resources when needed.
- It will report on progress to the senior management team. This should be done monthly because of the importance of the cross-functional strategies. The teams should also present progress during regular Hoshin plan reviews, which are usually done every 3 months.

Implementation plan

Hoshin plans may cascade down several layers of management. At the last layer, however, there needs to be an implementation plan. The implementation plan lists detailed steps or tactics necessary to accomplish the strategies in the Hoshin plan. Hence, implementation plans are typically prepared at the lower management levels or by professionals. Often a senior manager could have an implementation plan, if he or she has ownership of a strategy. This is the case in our example in Fig. 3.12.

The implementation plan format is shown in Fig. 3.11; the following is a discussion of each element. A completed example is given in Fig. 3.12.

Strategy with performance measure. Each strategy from the Hoshin plan is listed here, with the corresponding performance measure.

IMPLEMENTATION PLAN				
No	Strategy with performance measures	Implementation Details	Timeline Jan–Dec	Who

Figure 3.11

ANNUAL HOSHIN PLAN

| Prepared by: Quality Manager | Division/Department: Quality/APEX Division | Fiscal Year: 1989 | Page: 1 of 2 |

OBJECTIVE	NO.	STRATEGY (OWNER)	PERFORMANCE MEASURE
1.2 Improve Capability and structure of Quality Assurance throughout the computer system business.	1.21	Improve solution ratio of major customer complaints by setting up a Quality council (Quality Manager)	*Analyze and issue corrective action requests for all major complaints
	1.22	Establish formats and system for capturing customer quality problems for all computer products and be able to relay them to the production line quickly (Department Manager 1)	*Deviation from implementation plan *Data arrival at production line within 36 hours of customer complaint
TARGET/GOAL Goal #1: Obtain a solution ratio for major customer complaints of 90% Goal #2: Customer Satisfaction level improved by 10% in survey.	1.23	Reduce the recurrence of all complaints of customers (Quality Manager)	*Reduce recurrence rate by 50%
	1.24	All Dept. Managers to select one project for improvement (Dept. Managers)	*Project approved by Quality Manager *Project completed in 9 months
	1.25	Determine Training Needs (Dept Manager 2)	*Determine by March, 1989 *Provide Training by May 1989
	1.27		

Each strategy in the Annual Hoshin Plan should have a supporting Implementation plan, unless the strategy is to be deployed or cascaded further — for example Strategy 1.21. Here we show the implementation plan for strategy 1.23.

IMPLEMENTATION PLAN – 1989 (Quality Manager)

No	Strategy and Performance Measure	Implementation Details	Month												Who
			Jan	Feb	Mar	Apr	May	June	July	Aug	Sept	Oct	Nov	Dec	
1.23	**Strategy:** Reduce the recurrence of complaints from customers **Perf. Measure:** Reduce recurrence rate by 50%	• Set up task force • Set up Criteria for definition of major customer complaints • Analyze past complaints & reasons for recurrence • Validate reasons for recurrence • Prepare action plan & implement action plan • Check results, situation resolved? • Document procedure	X X		X		X X	X X X	X	X	X	X	X	X X	AB

Figure 3.12 Developing an implementation plan from a Hoshin plan.

Implementation details and timeline. Each strategy should be planned in detail. There are two choices:

- *Gant chart format.* List all the steps that need to be done for the strategy and list dates in the timeline.
- *PDCA format.* List all the steps in the PDCA cycle (see Chap. 4, "The Improvement Cycle.") In addition, list dates in the timeline. This is the recommended format.

We stress the need of a detailed implementation plan; the benefits include:

- Every Hoshin strategy in the organization eventually ends up in a detailed implementation plan, with an owner and a timeline.
- Progress can be measured regularly by checking against the planned timelines, and deviations can be easily spotted.
- This provides an excellent format for reviewing an employee's plans and the progress of those plans.

Who. Ownership of responsibility must be clearly defined to avoid confusion.

Daily Management, or business fundamentals, plan

As discussed earlier, Daily Management focuses on keeping the house in order, that is, maintaining the performance of day to day, routine, or repetitive processes. No special effort other than establishing goals, control limits, and a monitoring system are required. The prerequisite is that these processes are well understood because there is a wealth of experience and knowledge, which is documented. *Daily Management requires effective management of routine processes, discovering abnormalities or deviations, and preventing their recurrence.*

What are the processes that require routine management? We will discuss these soon, but first let us review the recommended planning format, which is shown in Fig. 3.13; an explanation of each element follows. A completed example is given in Fig. 3.14.

Item. List the item to be managed. The Daily Management plan has items that are well understood and documented. Little improvement is required for these items and the expectation is that all customers (internal or external) are happy with the performance of this item. There is little high-level management involvement in these activities.

The Planning Process 75

DAILY MANAGEMENT/BUSINESS FUNDAMENTAL PLAN					
No	Item	Goal and Control Limits	When Reviewed	Monitoring System or Data Source	Owner

Figure 3.13

DAILY MANAGEMENT/BUSINESS FUNDAMENTALS PLAN
Department: Quality   Fiscal Year: 1989   Page 2 of 2

| 7 | Facilitate Entity's Customer Satisfaction Model. |

DAILY MANAGEMENT/BUSINESS FUNDAMENTALS PLAN
Department: Quality   Fiscal Year: 1989   Page 1 of 2

No.	Item/Objective	Goal/Action Limit	When Reviewed	Monitoring System/Data Source	Owner
1	Expenses at or below target	● 100%, +0%, -15%	Monthly	Accounting	A,B
2	Product environmental testing and UL appovals	● On Schedule +0, -5 days	After each project	Department records	B
3	Quality Audits after production per audit procedure	● Minimum 2 per product	Monthly	Department records	C
4	Customer visits to collect data analyze and feedback to all	● 10 to 15 per Quarter	Quarterly	Department records	C
5	Ensure prototype testing and review during the design stage	● On Schedule +0, -3 days	After each project	Department records	B
6	Facilitate Project Postmortems for all products after release	● Completion of projects + 1 month	Quarterly	Department records	C

Figure 3.14 A Daily Management, or Business Fundamentals, plan. Shown here is one page of the quality manager's Daily Management/Business Fundamentals plan. Note that this lists routine items that are done on a daily basis. These are items for which there is a wealth of experience and knowledge, and all these items should have standards on how they are to be performed. Hence, the goals and limits are based on experience and no implementation plans are required because there are established procedures. Note: This quality department facilitates the entity's customer satisfaction model. This is indicated in page 2 of the Daily Management plan. More details of this are given in Table 2.2. In fact the entire Table 2.2 would appear on page 2 of this plan.

Goal and control limit. Each item will have a goal, and all goals must have control limits. When a process deviates outside of its control limits, the deviation must be analyzed and understood, and recurrence must be prevented. More on this later.

When reviewed. Here we list when this item is reviewed or checked.

Monitoring system or data source. This lists for each item where the information on performance is obtained.

Owner. Ownership of responsibility must be clearly defined to avoid confusion.

Guidelines for Daily Management plans. How many items should there be on a Daily Management plan? We recommend an average of 10 items for each department or manager. This is a guideline; the concern we have is that it will be difficult to monitor too many activities. Hence, priorities must be set. Some examples of items that are suitable are given below for design, manufacturing, sales, and service.

Items in a sales and service entity
- *The sales process.* The planned sales for each month will be supported by the sales process, which can be monitored and managed. This requires monitoring the following, details about which are given in Chap. 5, "Daily Process Management."

  - Sales funnel to measure the health of current and future orders.
  - Other key processes in sales and marketing, such as sales won/lost postmortems and marketing promotions.

- Capturing the customer's voice and needs during routine sales calls.
- Service contract management (planned calls, preventive maintenance calls, deviation from plans, etc.)
- Account management (management of major accounts).
- Customer satisfaction (via surveys, customer complaint and feedback system, etc.).
- Expenses.
- Employee morale.
- Training and education.
- Managing and tracking quality circle activity.

Items in a design and manufacturing entity. Note that more details on the first five items are given in Chap. 5.

- The production plan for each month will be supported by the manufacturing process. This requires a process assembly and quality checkpoints chart.
- Managing the customer satisfaction model (see Fig. 3.14 and Table 2.2).
- Capturing the customer's voice via various methodologies: QFD, routine visits, etc.
- Project postmortems (of newly released products).
- Product design for reliability (of new products).
- Expenses.
- Employee morale.
- Training and education.
- Managing and tracking quality circle activity.

Each of these items, with its goals and limits, will appear on the Daily Management plan.

Documentation of processes. Daily Management is management of routine processes that are well understood and documented. What is good documentation? For a customer complaint system, it is the type of documentation we showed in Chap. 2. Customer satisfaction surveys require a survey form and a flow chart of activities.

Other examples of appropriate documentation are given in Chap. 5; in that chapter we will discuss how to manage many of the key processes in a design, manufacturing, and sales company. These key processes will have performance measures that are captured and monitored on the Daily Management plan.

Managing abnormalities and deviations. Daily management also requires discovering abnormalities or deviations and preventing their recurrence. This is important and a system must be in place to allow for it. Good documentation helps. In addition, all deviations must be analyzed and understood, and recurrence must be prevented. There are two methods of doing this:

Use a Hoshin review table. This is illustrated and discussed in the section entitled "The review for Hoshin and Daily Management, or Business Fundamentals, plans." A review table is useful if deviations are reviewed every month or at some longer interval. Items requiring this method include expenses, sales quota, and customer satisfaction.

Use an out of control report. This is illustrated and discussed in Chap. 5 in the section on the Quality assurance system. It is useful for items that require frequent reviews—for example, a manufacturing process.

Control limits

Each goal in the plan should have control limits defining upper and lower bounds, for example, +10 and −10 percent. Figure 3.15 illustrates this concept. So long as values of the measure fluctuate within the limits, no investigation is necessary. If the limits, such as points A or B in the figure, are exceeded, causes for deviation should be uncovered and corrective action taken to bring the measure back within control in subsequent periods. *Remember that control limits are necessary for goals in the annual Hoshin and Daily Management plans.*

Ideally, control limits should be derived from data, which is used to construct a control chart. In such a case, we would have statistically computed upper and lower control limits. In an actual situation, however, this may be difficult, and the limits can be arbitrarily set. In this case, we would have limits that management sets based on experience or business needs.

Limits for Daily Management plans can be predicted because these are routine or repetitive activities. For annual Hoshin plans, because of the breakthrough nature of the objective, limits are more difficult to set—but management must agree on a minimum expectation. For example, if you wish to have a minimum sales goal of $1 billion, you could have a "want" target of $1.1 billion and a minimum acceptable "must" target of $1 billion. Setting limits has two purposes:

1. It is a recognition that goals are difficult to meet exactly, because there will be fluctuation in real life. For example, can you meet your expenses 100 percent as planned every month? Probably not.

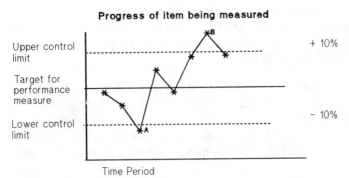

Figure 3.15 The concept of limits.

2. Management now has guidelines that show when to intervene in a situation: as long as the Daily Management item runs within limits, management need not worry. This concept is similar to control limits in statistical process control theory. There is more on this in the discussion on control charts in App. 3.

Setting numerical targets

Only breakthrough objectives—the key. For an organization to succeed and move onward, it is imperative that the number of objectives be limited—otherwise there will be dilution of effort and a lack of control. The Hoshin concept requires the selection of only breakthrough objectives. Management must resist the inclusion of numerous, often stereotype objectives, that are "perpetual virtues," or "motherhood and apple pie," or "hundred-year objectives"—that is, objectives that can be used year after year with little impact on the organization. The reasoning is that the breakthrough objectives will move the organization to the right position in the marketplace, while the Daily Management plan ensures that the house is kept in order.

Numbers game. Numerical targets have to be set for both objectives and performance measures. When this is done, the targets must be scientifically derived from facts and data. When setting targets for any item, you must look at the following information:

- What was the achievement over the last few years for this item and what success and difficulty have we had? What has been the average and what has been the range observed (lower and upper control limits)?
- What are other departments or entities achieving for this item?
- What have our major competitors achieved for this item?
- What are the customer's needs or requirements for this item?
- What are the major trends for this item in the industry and country?
- Do we wish to exceed our past performance or the competition for this item, and do we have the resources and technology to achieve the new target?
- What do our customers expect from us?
- What constitutes a breakthrough? Would it be a 10 percent change? Would it be a 30 percent change? You need to review the data to answer these questions.
- For guidelines about setting targets for improvement, refer to the discussion in Chap. 4 in a box entitled "Rate of Improvement and Setting Targets."

With the above information, you can set aggressive but realistic numerical targets. This gets you away from the numbers game, where numerical targets are set arbitrarily with dart boards or other equally ingenious means.

The review for Hoshin and Daily Management, or Business Fundamentals, plans

Basic methods of analysis. The annual planning process is not complete without comparing the actual results of the Hoshin plan with the targets. Any deviation between the actual and intended results must be analyzed in order to determine the reason for the deviation. The analysis should be carried out using statistical analysis of the data.

Even if the results are satisfactory—that is, performance is much better than plan—the process must be checked. This means determining the appropriateness of strategies and performance measures that were established at the beginning of the fiscal year. It is important to understand the reasons for success and failure—only then can we learn to do better.

This review is extremely important for two reasons. First, it is important to understand why you have met, exceeded, or fallen short of targets, and then it is necessary to take corrective action. Second, you must use this information—the lessons learned—to influence and shape future plans.

The review format is shown in Fig. 3.16; an explanation of each element follows. A completed example is shown in Fig. 3.17. Note that Fig. 3.16 is segmented into four elements, which represent the PDCA cycle.

Objectives and strategies. Each objective and its strategies from the Hoshin plan are listed and summarized here.

Actual. Actual results are listed for comparison with goals (for objectives) and performance measures (for strategies).

Status flag. This is not commonly used by companies that adopt Hoshin planning. We recommend the use of a status flag, a symbol similar to a traffic light. These are shown in Fig. 3.17, and an explanation of the symbols follows:

- *White circle (or on track).* To be used when a goal or performance measure has been met; if the goal or performance measure is planned to be accomplished at a later date, this symbol indicates that we are proceeding according to the implementation plan.

(Plan)	HOSHIN REVIEW TABLE		(Check)	(Act)	SS200
Objectives & Strategies	Goal or Performance Measure vs. Actuals	Status Flag	Analysis of Deviation	Implications On Next Planning Period	

Figure 3.16 Hoshin plan review table.

- *Cross-hatched circle (or warning).* To be used if there is a strong probability that a target or performance measure may not be achieved or if the implementation plan is not being carried out as planned. Management intervention may be required.
- *Black circle (or off-track).* To be used to indicate failure to meet target or performance measure. Management intervention is required.
- *P within a circle (or inappropriate strategy or performance measure.)* To be used when we discover that our strategy or measure is inappropriate. This often requires a mid-course change during the planning year.

Analysis. Causes for the observed difference between the Plan and Do phases are looked into. When there is deviation, it is important to understand the root cause—ask why five times. Remember to analyze both successes and failures.

Implications. The outcome of this phase will influence the Hoshin plan for the next planning period—this could be the next quarter (if reviews are done every 3 months) or next year (if this is the last review of the planning year).

Additional methods of analysis. Sometimes you may wish to provide a more detailed analysis than is provided with this format. This need may arise for technical problems, such as production yields. In such a case we suggest you use the Out Of Control report that we show in Chap. 5.

Guidelines for conducting reviews. Here are some hints for conducting an effective review.

HOSHIN PLAN REVIEW TABLE – 3Q1989

STATUS : ○ ON TRACK ◈ WARNING ● OFF TRACK (P) GOAL, STRATEGY OR PERFORMANCE MEASURE IS INAPPROPRIATE

Prepared by: Quality Manager Date: August 1989 Year: FY 1989 Div: APEX Location: Quality Department

OBJECTIVE/STRATEGY (P)	ACTUAL PERFORMANCE & LIMITS (D)	STATUS FLAG	SUMMARY OF ANALYSIS OF DEVIATIONS (C)	IMPLICATIONS FOR FUTURE (A)
Objective: 1.2 Improve capability and structure of Quality Assurance throughout the system business	**Goal #1:** 90% solution ratio Actual: 100%	○		• Solution ratio better than targeted due to effectiveness of Quality Council
	Goal #2: Customer satisfaction improved by 10% Actual: 5%	● (P)	Goal of 10% was too high. Experience has shown 7% is best we can do. Also one customer issue was not resolved because marketing manager quit.	• Set targets after analyzing past survey data. **Action:** General Manager • One customer issue not resolved (availability of self diagnostic tapes) because Marketing manager quit and no one else followed through. In future, Council must follow up on corrective action with all managers. **Action:** Quality Manager
Strategies: 1.21 Improve solution ratio of major customer complaints by setting up Quality Council	**Performance Measure:** Analyze and issue corrective action for all major complaints. Actual: 10 out of 13	◈	3 of 13 are in process. Will expedite this month.	• No further action needed. The regular meeting of Quality Council has helped ensure every customer issue is resolved.
1.22 Establish formats and systems for capturing customer complaints	**Performance Measure: #1** Deviation from plan Actual: Met plan	○		• Doing very well. In future we need to look at how quickly we resolve customer complaints. We will select this as an item for improvement next year.
	Performance Measure: #2 Data at line within 36 hours. Actual: Average to date 34 hours	○		

Figure 3.17 A Hoshin plan review format. This figure shows a review of the quality manager's annual Hoshin plan which was shown in Fig. 3.3. Note that both the objectives and strategies are reviewed. Also note the status flags, or "traffic lights," which give a quick and precise indication of progress to plan.

Frequency. For an entity or function we suggest quarterly (3-month) reviews for the Hoshin plan. For the Implementation plan, a monthly review is recommended.

Format. We are proposing a standard format. The same standard format is used for quarterly and annual Hoshin plan reviews and also for Daily Management, or Business Fundamentals, plan reviews. The format is shown in Figs. 3.16 and 3.17. Note the use of status flags to indicate if the goals and performance measures are on track, off track, and so on.

The symbols are very useful to give quick, visual feedback on the status of the plans. For example, if there is a black circle to show that the objective's goal is not being met and some of the performance measures are white circles, it will be obvious that the objectives and strategies don't match.

Review procedure
- Discuss objectives and strategies.
- Review goals and performance measures, that is, compare actual performance against target.
- If there is deviation, state the reasons for deviation in the analysis column and list countermeasures in implications column.
- If progressing to target or met target, put comments and give learning points in implication column. Remember, the review table is in PDCA format and the cycle must be completed and documented.

For fourth-quarter and annual reviews
- Use the same procedure and generate issues that affect next year's plans.
- Typically, the third-quarter review would become the annual review. For this we suggest you do the following: Review the actual performance up to the third quarter and provide a prediction for the fourth quarter. The prediction should be based on data and trends. This review will give you inputs for the next year's plan.

Duration of review. Each manager could take up to 60 minutes for his or her presentation. Therefore, general managers and their staffs will take about 1 day for a Hoshin review.

Some additional review pointers—what to look for
- The manager under review should be concise. In the analysis column put the facts and root causes. And in the implications column indicate the corrective action that is planned.
- The manager under review should have backup data, especially because the review format requires a summary of data.

- Focus on failure to meet plan rather than on success. But if there are good reasons for success, they should be highlighted, understood, and disseminated.
- If a target is scheduled for, say, the fourth quarter of the planning year, it will be insufficient to say that you are on track during the early part of the year. The danger signals that may show up are as follows:

  Comment during the first quarter: OK, on track.

  Comment during the second quarter: OK, on track.

  Comment during the third quarter: OK, expect to meet goal.

  Comment during the fourth quarter: Oops! problem. No resources. Missed goal.

Therefore you must check progress by reviewing the milestones, implementation plan, Gant chart, etc. Check these; if they are not available, there may be a problem. Be especially wary of comments such as no problem or on track. Always check the facts, data, or milestones.

Changing objectives or strategies during the year. Often, an entity may discover that an objective or strategy is wrong or inappropriate. Other times, there may be external factors requiring changes in an objective or strategies. Making changes in the plan is therefore appropriate and recommended.

Reviews of implementation plans. For the implementation plan we recommend a monthly review. No special format is recommended. The review can be done between the owner of the implementation plan and the immediate supervisor. At this point, progress to plan is checked. Any deviation or slip should be analyzed and corrected. Notes or comments can be written directly on this plan. These plans can also be presented during a quarterly Hoshin review to support any statements or comments. For example, if you are on track with your plan, support the statement with your progress with the implementation plan.

Hoshin plan reviews are done bottoms up. A final word on Hoshin reviews. They should be done bottom up. That is, first the implementation plans are reviewed at the lower level, followed by reviews of higher-level plans.

Reviews of implementation plans. Implementation plans are reviewed and then the Hoshin plans that drove the implementation plans are reviewed; then Hoshin plans for the general manager's staff are reviewed; after that the chief executive can review the Hoshin plans for several product or sales divisions or plans for general managers. This process is necessary—it should take less than 4 weeks—to ensure a thorough review, with necessary corrections, and to keep the ship going in the right direction.

Planning: Putting It All Together

Let's collect our thoughts and put together the ideas we have discussed on the planning process. It begins with the company's or organization's vision and continues with the long-term plan, the annual Hoshin plan, the Daily Management plan, and the implementation plan and finishes with regular reviews of progress. Then the entire cycle starts again. We illustrate this in Fig. 3.18.

Figure 3.18 summarizes all the requisite activities that are required for a specific department or entity. A suggested timeline is also given. The most crucial and difficult portion will be the implementation plan. This is the point where all the best-laid plans must be implemented and culminate in results. We recommend you give special attention to this area.

Also, remember that all the various tactical plans or projects must be assigned to specific managers, professionals, cross-functional teams, or quality teams. Where possible, these should be the only things they are working on in addition to their routine or daily assignments (Daily Management activity). This will ensure tight linkage from higher-level plans down to each individual person in the organization and will result in a single-minded approach by the entire team to achieve objectives.

Planning calendar

In Fig. 3.18, we give a suggested timeline for all the activities. It is important that you prepare a planning calendar for your entity for all the various activities. This should indicate activity versus responsible person or department. It also sets expectations for the planning process each year.

Planning for special or R&D projects

You may wonder where any special, ad hoc, or R&D projects fit into this planning methodology. Since we feel that all activity must be planned, documented, and reviewed, we suggest the following: For critical projects or technology breakthroughs, the project obviously goes into the Hoshin plan; otherwise the projects can be on the appropriate department's Daily Management, or Business Fundamentals, plan or on a separate project list.

86 Chapter Three

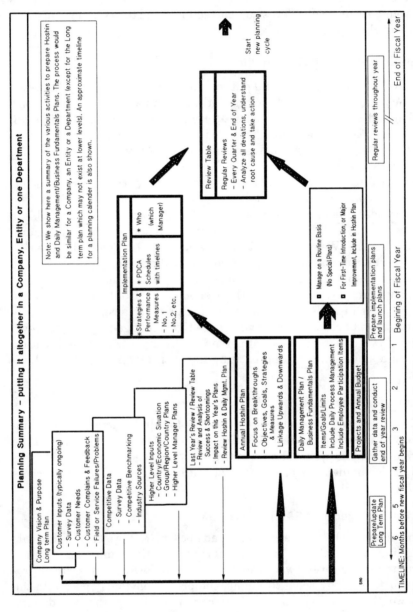

Figure 3.18 Planning summary: Putting it all together.

Planning and budgeting

The annual budget and sales targets must be tied closely to the planning process. Once the long-term and annual plans are completed, the annual budget must be finalized, although preparation can start during the planning stage. It is important that the annual budget is not just a tweaking or simply an increase of the previous year's budget. Instead, it must support the upcoming year's plans. Hence, it is crucial to prepare the final budget after the plans are done; otherwise plans and resources (people and dollars) may not match.

Note: We feel that our recommended planning process must be done in its entirety; otherwise do not do it at all.

Questions and Answers on Hoshin and Daily Management, or Business Fundamentals, Planning

Many organizations that start to use Hoshin planning do not reap its benefits immediately. There are several reasons for this, and we will cover them through a question and answer session:

Q1. What are the one or two most common problems seen when an organization starts to use Hoshin planning?

A1. *One of the most common problems is that of listing too many objectives at the higher levels of management.* This will cause a high fan out or cascading to lower levels. The result is a vast and unmanageable plan and a loss of focus for the organization. The key is to have a few breakthrough objectives at the product division or entity manager's level. At the CEO or senior manager's level, *one or two breakthrough objectives are probably all that can be managed well.* In addition, senior manager's need to avoid setting 100-year objectives.

Q2. Hundred-year objectives—what are they?

A2. That means an objective that can be used every year for the next 100 years. It is what many Americans call motherhood and apple pie. Examples are objectives such as "understand our customers better," "achieve market success," or "work smarter and harder." These are statements that mean little. What you must have are specific objectives with measurable goals.

Q3. Sometimes there is little choice except to have too many objectives. For example, in a matrix-type organization, a high-level manager may have several bosses. Or just any manager may get several requests from superiors via their Hoshin plan or other inputs. How can you resolve this without generating five, six, or even ten objectives?

A3. Yes, this situation may occur. You have two choices that can help you minimize this problem. One is to group several requests into one or more "super" objectives. For example, you may get requests to reduce costs, increase productivity, and reduce failure rates. These three items could become three strategies in one objective: to increase profits (see the example in Fig. 3.5). This way you have avoided getting three separate objectives plus maybe 15 strategies at your level. The trick is to be as strategic as possible.

The other choice is to delegate some issues or requests that appear at your level, especially if these are too tactical. Or you can try both choices. What all this means is that not every issue or request (from a higher level) needs to become an objective at your level. This way you are able to control the number of objectives at your level.

Q4. What advice do you have on the use of a Daily Management, or Business Fundamentals, plan?

A4. There are four common mistakes you should avoid:

- The list should be kept short. Senior manager's should have a short list—perhaps 10 items. A longer list, several pages long, can result in a loss of focus and unnecessary generation of paper. In manufacturing or sales, at the lower levels, one of the items could be a list for managing the manufacturing or sales process.
- Keep Hoshin and Daily Management plans separate. A common problem is to have both types of items on a Hoshin plan. This mistake generates too many objectives, causing clutter, and can result in improper objectives and goals—specifically, a lack of breakthrough items.
- A third mistake is the difficulty of setting goals for Daily Management plans. We recommend you set these goals based on experience. When you have no experience in a particular item, it may be wiser to make it a Hoshin plan strategy for the first year. For example, "set up and measure a process" may be the first-year strategy. In the second year, this process may appear in the Daily Management plan, with a data-based goal.
- A fourth mistake is the lack of limits for all the goals (refer to the section "Control Limits" earlier in the chapter).

Q5. I have read that management consultant Tom Peters does not believe in planning. In fact he suggests that you don't need plans and goals.[4] Can you comment?

A5. We have a tremendous amount of respect for Tom Peters. Typically, he brings fresh insight into management methods and thinking. We think he was against slow decision making, which is often cloaked in the guise of planning. If a manager has some data and continues to delay making decisions, we agree with Tom Peters that this is an unforgivable sin.

On the other hand, in contrast to his numerous studies showing that planning is not helpful, we can quote studies indicating that planning helps tremendously. The successful Japanese companies are great planners—you only need to look at Matsushita, Toyota, and NEC. Nevertheless there must be a balance. Too much planning is bad, and we have recommended short plans with only a few objectives. In addition, when necessary, quick decisions are a must. Many successful products have been the result of a quick management decision in the absence of sufficient data.

Q6. How do you convince a skeptical manager to adopt the Hoshin planning methodology?

A6. There are two answers to this. First, we hope that your company adopts this as a standard to replace a weaker planning process. In such a case, our question to you is why isn't the manager adopting the company standard? Second, your other choice is to analyze the current planning process. For example:

- What is the current process?
- How are current objectives deployed down the organization and are they linked to the general manager's objectives?
- How are current objectives on the plan reviewed?
- What were last year's plans; what was achieved; what was not achieved? Were there problems?

If you discover enough weaknesses, it will be easier to convey the benefits of the Hoshin planning methodology.

Q7. Must we have a long-term plan, in addition to the annual Hoshin plan?

A7. Not always. Most departments and operations will not need such a plan. But each organization—say, a product division or a sales division or headquarters—should have a long-term plan that addresses issues and strategies over a 3- to 5-year time frame.

Q8. What about implementation plans?

A8. Remember to have implementation plans. At a certain level in the organization, specifically the professional level (engineers, accountants, supervisors, sales representatives, etc.), you should only have implementation plans.

Q9. I have heard that one of the best ways for a senior manager to be involved in TQC is via Hoshin planning. Any comments?

A9. Yes, that is a valid statement. If you look at Fig. 3.18, you can see that the Hoshin plan requires addressing the entity purpose and vision, customer needs and issues, quality, costs, process management, employee participation, and so on. These are all elements of TQC, and Hoshin planning ensures that you address them—if you follow the process correctly.

Q10. Can you have the same item showing up on the Hoshin plan and the Daily Management plan?

A10. No. You should either be maintaining (or slightly improving) an item on the Daily Management plan or having a breakthrough improvement on the Hoshin plan. But on a year to year basis an item can move from one plan to the other. For example, product reliability could be on the Daily Management plan in one year, but due to a catastrophe, reliability could become a problem. Then in the following year, the operations manager could have a Hoshin plan objective to improve product reliability.

Q11. Any other words of advice?

A11. Yes. Don't get carried away with the Hoshin process. Keep the plans short and concise. Senior managers should try to keep to two breakthrough objectives and a short—one page—list of business fundamentals or Daily Management items. That gives up to three or four pages using the format we have given you. If your plans are longer, they may be appropriate, but we suggest that you look again at your priorities.

Summary: The Planning Process

Planning is one of the most important processes in an organization—some would say the most important. It is what drives most other activities.

We have proposed the use of both long-term and annual planning methodologies. For the annual planning process we have recommended the use of the Hoshin and Daily Management, or Business Fundamentals, plans. The Hoshin plan focuses on the organization's breakthrough objectives, while the Daily Management plan is used to manage the organization's day to day activities; more information on Daily Management is provided in Chap. 5.

Hoshin and Daily Management planning are systematic and tightly coupled processes. They require effort and consensus and in return provide a focus, a single-minded approach by the entire management team. The process is designed to enhance the chances of success.

During planning, we recommend that you review four crucial areas: quality, costs, delivery, and education. This is to ensure that there in no neglect in these important, generic success factors.

During the planning process, you will discover many items that need improvement. In the next chapter we will discuss how to manage the improvement cycle.

Finally, remember that formal planning provides many benefits, including systematic thinking, better coordination, sharper objectives, improved performance standards, and management involvement. All of these result in a planned approach to tackling the marketplace that eventually can end in higher sales and profits.

References

1. Soin, Sarv. Singh. Much of the Hoshin planning methodology mentioned here comes from the author's book, *TQC at Hewlett-Packard—The Asian Experience,* second edition, an internal publication. Additions and changes have been made based on the author's personal experience. The categories (objectives, goals, strategies, and performance measures) that we show are similar at companies that use Hoshin plan methodology in the United States and Japan. Their forms, however, may be different.
2. Merrills, Roy. "How Northern Telecom Competes on Time." *Harvard Business Review,* July/Aug. 1990, pp. 108–114.
3. Ames, B. Charles, and James D. Hlavecek. "Vital Truths about Managing Your Costs." *Harvard Business Review,* Jan./Feb. 1990, pp. 140–147.
4. Peters, Thomas J. "Want to Get Ahead? Then Don't Plan, Do It." *San Jose Mercury News,* Dec. 3, 1990.

Chapter 4

The Improvement Cycle

*We need never ending improvement . . .
to establish better economy.*
                    W. EDWARD DEMING

Overview

One of the basic tenets of TQC is that of continuous improvement. The old adage "don't fix it if it isn't broken" does not work anymore. In today's environment, if you don't fix it, your competitor will and in addition will take away your market share. Just look at the business that express mail companies have taken away from the U.S. post office or look at what the Japanese electronics industry has done to their European and American competitors.

We also need to improve things in a systematic way. There are numerous tools available that will help, such as the PDCA improvement cycle—on which we will devote much time—design of experiments, and Taguchi methods. But before we go any further, here is an interesting incident:

> I visited a R&D lab recently, and they showed me the various integrated circuit products that they design. There was one interesting driver assembly that they supplied to a display manufacturer. They told me that this current model was the third-version assembly. What was different between this and the first version? The first version had some failings, so they decided to improve it. The second "improved" version was short lived because it did not adequately resolve the failings of the first version, so they spent several more months developing the third version.
>
> But wait, why did the second version not solve the problems? Well, the R&D manager admitted, they thought they had the fix in the second version, but they were wrong. Thought! . . . that's right, thought—they used gut feel and experience to arrive at the solution—little analysis had been done.

Alas, this is a common tale that I have seen on numerous occasions: the use of experience and genius to solve problems. This is little different from Russian roulette, except that it is the customer of your company who gets shot.

Selecting Items to Improve

How do we select items to improve? Every organization will have a poor quality "iceberg" of visible and hidden problems. We need to know these problems. Figure 4.1 shows a poor quality iceberg contributed by Mike Ward of Hewlett-Packard. In it we list some of the obvious and hidden problems that could occur in a large company.

It is extremely dangerous to ignore or hide these problems. Consider the following statement:

> I am sick and tired of visiting plants to hear nothing but great things about quality and cycle time—and then to visit customers who tell me of problems.—*John Akers, Chairman, IBM Company,* Newsweek: *June 10, 1991.*

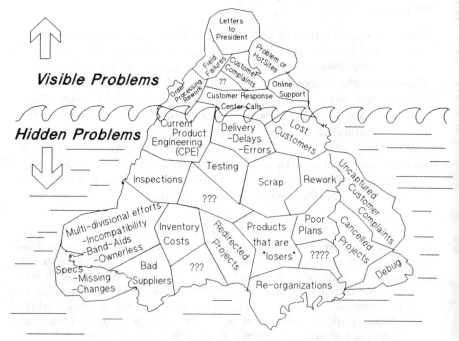

Figure 4.1 The poor quality iceberg. This figure shows the iceberg of problems that may exist at an organization. Typically, we may be attacking the obvious problems that exist, not realizing the numerous hidden problems or opportunities for improvement. These opportunities, if left unresolved, can lead to waste, higher costs, and customer dissatisfaction, resulting in lost business. A well-managed organization should have a small iceberg.

Sound familiar? The solution is to have a system in place that captures and resolves these problems. Throughout this text we provide methodologies for identifying problems. Briefly, these include:

Chapter 2, "Customer Obsession." The various customer satisfaction surveys that are conducted, competitive analysis and benchmarking, the customer complaint and feedback system, the product satisfaction model, and of course product and service failure information systems

Chapter 3, "The Planning Process." All the issues raised during the preparation of the long-term and annual Hoshin plans and items discovered during Hoshin and Daily Management plan reviews

Chapter 5, "Daily Process Management." All the processes that need to be managed and improved upon

Chapter 6, "Employee Participation." All the issues raised by employees and during preparation of the entity's quality direction

The above methodologies will provide fertile ground for identifying items for improvement. Before we start improving, priorities must be set; otherwise we may have too many things to do. This is done routinely during Hoshin planning, discussed in Chap. 3, and via the quality assurance system, discussed in Chap. 5.

Cost of Poor Quality

The various items shown in the poor quality iceberg can be itemized, grouped, and converted into dollars wasted. This method is used by many consultants in America and Europe. It was popularized by A. V. Feigenbaum[1] and Joseph Juran[2] and called quality costs. Others call it cost of quality, which is not appropriate—a better description is the cost of poor quality.

Juran talks about three types of costs: internal failure costs, external failure costs, and prevention costs. He then goes on to describe a model that gives "optimum quality costs." He states that these costs can be made to decline on a continual basis. Certainly, this is one way to get management's attention to start a quality improvement program—that is, by focusing on dollars wasted. In many companies this is the only way to get management's attention. This is very different from the approach adopted by Japan and the newly industrialized countries. Here quality is improved because there is a strong drive toward perfection and customer satisfaction, resulting in increased market share and profits.

The notion of managing quality costs is disputed by Hitoshi Kume.[3] He contends that Western companies "are so concerned about identifying quality costs that there seems to be an impression that quality control activities cannot exist without a formal quality cost system." He goes on to say that he has tried and failed to introduce this concept in Japan—partly because it is impossible to include all of a company's quality information in quality costs. He gives several examples to support his argument. For example, a mature product (like a TV set) will have low quality costs and low profits—because of high competition. Whereas a new, innovative product could have high quality costs and yet have high profits—because of no competition and a high retail price. He goes on to say: "The goal of business management is to increase profit, not to reduce cost." Therefore cost increases are acceptable, so long as a company can obtain more profit to offset the additional cost.

The most important thing that management needs to do is to ensure that the design, production, marketing, and product meet the customer's needs. Hence, although cost should be lowered on a continual basis, lower quality costs are not necessarily a sign of successful management. If all costs are equal with other firms, success can only come from continuous development and introduction of new products that meet customer needs—because the largest loss is probably the loss of market share, and quality costs do not measure this. In fact in the same vein Edward Deming has said: "The most important costs are unknown and unknowable."

In summary, we do not recommend the laborious collection of cost of poor quality data. Nevertheless, this may be useful in a specific department that has very high poor quality costs. Instead we suggest using the guidelines given in the section "Selecting Items to Improve." A good management system will automatically and continuously focus on improvements. Concurrently the focus must be on developing products and services that meet and exceed customer needs.

PDCA Cycle

At a seminar I once attended, a participant asked the legendary Kaoru Ishikawa ,"What is the most important tenet of TQC?" He replied without hesitation, "The PDCA cycle."

The PDCA (Plan Do Check Act) cycle was originally developed by Walter Shewhart—the originator of Statistical Quality Control. It was popularized by Edward Deming[4] and is often called the Deming cycle. It gained wide popularity in Japan, through the efforts of Deming. Since then there have been numerous versions called the TQC story, the QC story, CA-PDCA, ad infinitum. Very briefly, here is an explanation of some of the versions.

The Shewhart and Deming cycle

This is a cycle designed to help in improving a process. It is also meant to be used as a procedure for finding the root cause through statistical analysis. It is divided into four steps, as follows:

1. What is to be accomplished? What data are available? Are new observations needed? If so, plan and decide how to get more data.
2. Carry out the change to be accomplished, preferably on a small scale.
3. Observe the effects of the change.
4. Study the results; what can we learn or predict?

The PDCA cycle

The PDCA cycle is very similar to the Deming cycle. The four words *Plan Do Check Act* describe the stages very nicely and are more explicitly stated as follows:

1. *Plan.* Determine goals and methods to reach the goals.
2. *Do.* Educate employees and implement the change.
3. *Check.* Check the effects of the change. Have the goals been achieved? If not, return to the Plan stage.
4. *Act.* Take appropriate action to institutionalize the change.

The CA-PDCA cycle

The thinking behind the CA-PDCA cycle is that you need to check or analyze the current situation before you start to Plan, Do, Check, Act. The logic behind this is correct, but why not just add a step in the plan that requires analysis? That was Shewhart's original intent. Doing this will allow the original PDCA cycle to be retained.

QC story

This attempts to cut through the confusion of the various improvement cycles and provides a sequence of activities similar to the CA-PDCA cycle, without using the words *Plan Do Check Act*. A word of caution on the QC story: Many people have the impression that the QC story is only meant to document a project when it is completed. *This is wrong.* It is meant to be used as a step by step guide for solving a problem and as a procedure to document a completed project. The same concept applies to the PDCA cycle that we will now discuss in detail.

The PDCA Cycle

Start here → **1. Select the theme or project**
- plan the schedule of activities
- set the target

7. Conclusion and future plans
- Continue on same issue or select new issue

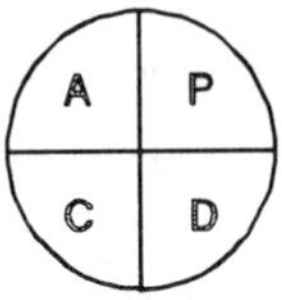

2. Grasp the present status
- Get and review the data

6. Take appropriate action
- Standardize, control and document
- Train and educate

3. Analyze the cause and determine corrective action
- Cause & Effect
- Set Hypothesis
- Verify most likely causes
- Determine corrective action
  - short term or remedial
  - long term or preventive

```
  A | P
  -----
  C | D
```

5. Check the effects
- Compare the results to the target
- Go to Plan if target is not achieved

4. Implement corrrective action
- Take corrective action
- Conduct adequate training

Figure 4.2

Modified and improved PDCA cycle

Figure 4.2 shows a modified PDCA cycle, which retains the original intent of the cycle but includes the various improvements of the other versions. This is the cycle we recommend to you.

The PDCA cycle is often depicted as a wheel, as shown in the center of Fig. 4.2. This is an important concept because a turn of the wheel represents one improvement cycle, which brings us to the beginning of the next cycle. When one cycle is completed, there are two alternatives that can be pursued: control the improved process or go through another improvement cycle.

Relationship between Improvement and Control

Let us look at the relationship between improving and controlling something. As previously stated, at the end of an improvement cycle we have two choices: put the improved process under control or start another improvement cycle, after putting it under control. We illustrate this concept in Fig. 4.3. The choice is influenced by the nature of the current project and other priorities. The purpose of putting it under control is to maintain the improvements that have been made—because it is very easy to slip into old habits and lose the gains. Hence, proper training and documentation is crucial to help retain the gains.

The Improvement Cycle 99

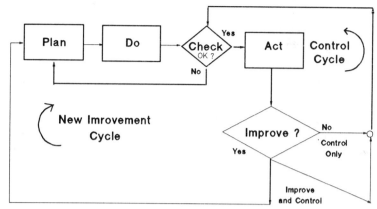

Figure 4.3 The relationship between control and improvement.

In the next chapter, we will discuss process control and management of processes.

The alternative is to go into another cycle of improvement, after the process is put under control—if not now, later. At that point good documentation of the current project—the analysis, the validation, the choices made, the gains, and what remains to be improved—is very important. Having such information will make the next improvement cycle easier and faster.

Benefits of the PDCA Improvement Cycle

Here are the main benefits of the PDCA improvement cycle:

- It is a systematic problem-solving process that provides the quickest route to an effective solution.
- It ensures an agreed upon schedule for project completion.
- It ensures an agreed upon goal or target, usually set with data.
- It ensures a detailed analysis of the failure modes.
- It ensures verification and elimination of the most likely failures modes.
- It requires implementation of controls to monitor and manage the new improved process.
- It requires training in and documentation of the new process.
- It requires documentation of before and after fail data. This will be useful for the next improvement cycle.

- It will ensure no recurrence of the problem, thus ensuring continuous improvement. This is achieved through standardization of the new improved process.
- Managers and supervisors may come and go, but if PDCA is institutionalized and mandatory, employees will always be systematic and analytical in eliminating root causes of problem areas.

The last point is important. We paraphrase it as follows: People may come and go, but processes stay. The PDCA cycle is part of a process—the improvement process. Over the years, we have seen numerous improvement processes, each customized to meet an individual's needs or pet theory. This causes an excessive amount of relearning and inefficiency.

We suggest you minimize the use of other improvement processes. Instead, standardize the use of the PDCA process shown here. This will result in a common language and facilitate communication of improvements in an organization. Many companies have adopted this as a standard, including most Japanese companies, and in the United States, Florida Power & Light and Hewlett-Packard.

Detailed PDCA cycle

Shown next is a detailed PDCA cycle. You will find it useful as a reference and a training tool. Each of the seven steps in Fig. 4.2 is explained in greater detail. In addition, we have listed some quality control (QC) tools that could be used in each of the steps. To illustrate the steps in the PDCA cycle, we give a true example, which comes from Hewlett-Packard Malaysia.

Plan stage

Step 1: Select the theme or project
- *Objective of this step.* To clearly define the problem to be resolved.
- *Discussion.* Here we define the project, understand its background, set a target, and prepare a schedule of activities. The following are substeps of step 1.

Step 1a: Project background and reasons for selection. The project can be selected from the department objectives and customer complaints, or it can be a continuation of a previous project. And of course it should be within the improvement team's control.

Step 1b: Set a target. This should include a statement of the item, a numerical number to be achieved, and a time frame for project completion, for example, Improve product yield to 85 percent by May 1991. Set reasonable and realistic targets; otherwise, it will be difficult to achieve them. The target can also be set after step 3, when more data is available. The situation will vary with each project, but an initial target may be appropriately set here. The target can be set using these guidelines: Use current data to set a breakthrough goal, use competitive data to equal or better the competition, or use the rule of thumb of reducing defects by 50 percent every 12 months. Refer to the rule of thumb given later in this chapter in the box "Rate of Improvement and Setting Targets."

Step 1c: Prepare a schedule of activities. This lists the seven steps in the PDCA cycle and the expected time frame for each step. For our first project we may estimate, and with experience this will be easier. We recommend the implementation plan format, shown in Chap. 3, "Planning Process," to record the schedule. This is very important since it sets outer bounds on the project.

- *QC tools that can be useful.* Pareto diagram, trend charts.

Step 2. Grasp the current status
- *Objective of this step.* To understand the problem area and to highlight specific problems.
- *Discussion.* Here we study the effects of the problem by reviewing the available data. Our study should be approached from several facets, such as time, location, and type. For example, if we want to reduce the percentage failures in a facsimile machine, for time, we can look at the failures between the day and night shift and over a period of time. For location, we can look at which part of the machine

---

Example of Step 1, Project Theme

Step 1a: Project background and reasons for selection

To improve first-pass yield at final test for product QDSP-6666. This product is a small display, used in cars and computers.

1. Low first-pass yield at final test (86.2 percent)
2. Too much rework at final test
3. Shipment timeliness adversely affected

Step 1b: Set a target

To achieve 96 percent first-pass yield at final test by July 1988. This is equivalent to reducing failures from 13.8 to 4 percent and is in line with the department goal.

Step 1c: Prepare a schedule of activities. This is not shown here.

has most failures, for example, top, bottom, side, center, etc. For type, we can determine which aspect of the machine is causing failures, for example, the printed circuit board, power supply, or plastic case. The available data can be presented in graphs and Pareto charts. We should get a process flow chart of the product or process being studied. If it does not exist, we must prepare a chart.

- *QC tools that can be useful.* Process flow charts, Pareto charts, trend charts, control charts, histograms, process capability indices.

Step 3. Analyze the cause and determine corrective action
- *Objective of this step.* To find out the root cause of the problem and to plan for corrective action.

- *Discussion.* In this step we examine the causes of the problem, isolate the root causes, and determine corrective action. The detailed substeps (3a, 3b, and 3c) follow:

Step 3a: Prepare cause and effect diagram. The item to examine is selected. This can be the first or second bar in a Pareto diagram of defects, or it can be the first two or three bars. Other times we may select a specific item we want to improve. A cause and effect diagram is then prepared. All the causes in the cause and effect diagram can be derived through a brainstorming session. Possible and impossible causes should be listed. Now use data that was obtained in step 2 to eliminate unlikely causes. The cause and effect diagram can be simplified and redrawn.

- *QC tools that can be useful.* Pareto diagram, cause and effect diagram.

Example of Step 2, Current Status

In Fig. 4.4, we show a graph and Pareto diagram. The graph gives the final test yield data for the display assembly, and the Pareto diagram gives a breakdown of the causes of failures. A process flow chart is available but not shown here. Note the following for the Pareto diagram categories: Open Digits means any digit that does not light up. Wrong sequence means the displays fail to exhibit the correct sequence as per the product specifications.

The Improvement Cycle 103

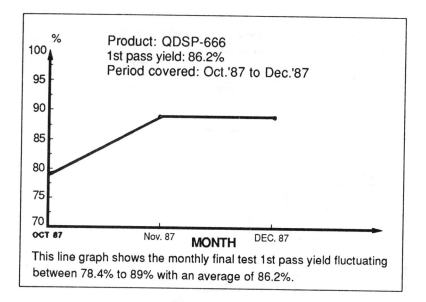

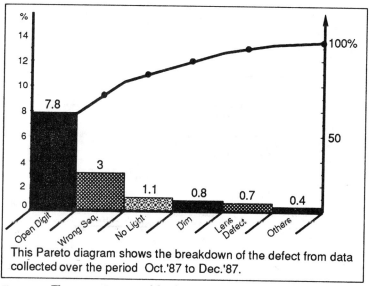

Figure 4.4 The current status of display assembly failures.

Example of Step 3, Analyze the Cause and Determine Corrective Action

Step 3a: Prepare cause and effect diagram

The two highest fail categories in the Pareto diagram—Open Digits and Wrong Sequence—were selected and a cause and effect diagram was prepared for each item. The group members then brainstormed for all the possible causes of each item by using the questioning techniques of the 4Ws (what, where, when, why) and 1H (how). The cause and effect diagrams are shown in Fig. 4.5.

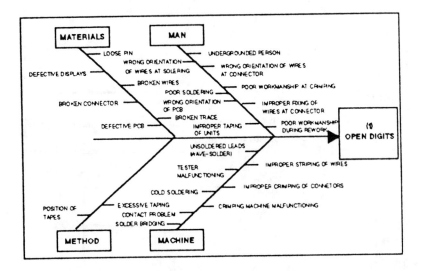

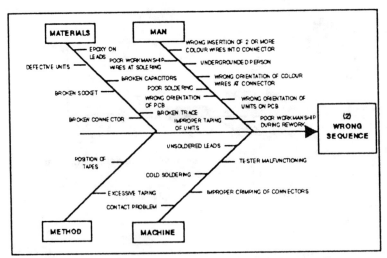

Figure 4.5 Cause and effect diagrams for Open Digits and Wrong Sequence.

Step 3b: Prepare a hypothesis and verify most likely cause

- *Discussion.* We prepare a hypothesis by selecting the most likely causes out of the cause and effect diagram; this can be done using the group's experience or voting. This list of most likely causes must now be verified with data. But we should use new data to determine if there is a relationship between the selected causes and the effect; this may require us to conduct experiments. We will now have a short list of verified causes—the root causes of the problem we want to reduce or eliminate.

- *QC tools that can be useful.* Checksheet, stratification, statistical design of experiments.

Example of Step 3b, Prepare a Hypothesis and Verify Most Likely Cause

Members deliberated on the possible causes and then voted for the most probable ones. The basis for the selection was the operators' job experience and knowledge. They are as follows:

1. Open Digits
   a.) Wrong orientation of wires inserted into connector
   b.) Improper fixing of wires into connector
   c.) Broken wires
   d.) Unsoldered leads
   e.) Defective units
2. Wrong Sequence
   a.) Wrong orientation of wires inserted into connector
   b.) Wrong orientation of wires at soldering
   c.) Improper crimping of connectors

Verification

Additional failure data was collected in order to verify which of the possible causes were the most likely ones. The analysis findings were then summarized as shown in Fig. 4.6. Members also noticed that most of the causes were workmanship related. The most probable causes were verified to be:

1. Wrong orientation of wires inserted into connector
2. Wrong orientation of wires at soldering
3. Improper fixing of wires to connector
4. Unsoldered leads

TYPE OF DEFECT	Open Digit	Wrong Sequence	Open Digit & Wrong Seq	Total
Wrong orientation of wires at connector	16 (53.3%)	13 (76.5%)	4 (100%)	33 (64.7%)
Wrong orientation of wires at soldering	4 (13.5%)	4 (23.5%)		8 (15.7%)
Improper fixing of wires to connector (Wires came out loose)	3 (10.0%)			3 (5.9%)
Unsoldered leads	4 (13.3%)			4 (7.8%)
Broken wires	1 (3.3%)			1 (1.96%)
Defective unit	1 (3.3%)			1 (1.96%)
Improper crimping of connector	1 (3.3%)			1 (1.96)
QUANTITY ANALYZED	30 Sets	17 Sets	4 Sets	51 Sets

Figure 4.6 Verification of most likely causes.

Step 3c: Determine corrective action

- *Discussion.* We decide on the corrective action. Sometimes the corrective action is obvious. If not, we need to decide on the action. Creative alternatives should be generated using brainstorming or cause and effect diagrams. There will usually be two types of corrective action:

  - A quick fix or remedial action. This could include inspection for the defect or repair of the defect.

  - A long-term fix or preventive action. This could include elimination of the root cause, hence preventing the problem from recurring. This is more important, but because of constraints, the quick fix may be implemented first.

  It may be necessary to conduct a trial of the proposed action to determine that it works. Only then should it be proposed.

- *QC tools that can be useful.* Checksheet, checklist.

Do stage

Step 4. Implement corrective action

- *Objective of this step.* Implement the plan and eliminate the root causes of the problem.

- *Discussion.* Employees who execute the correction must understand the corrective action. Good communication and training will be necessary. The following are recommended substeps.

The Improvement Cycle 107

> **Example of Step 3c, Determine Corrective Action**
>
> Since the four major causes were workmanship related, the members attempted to list the snags and difficulties in the processes that are related to each of the causes. Members then suggested and discussed what could be good solutions and/or preventive actions for each of these. The discussion is summarized in Fig. 4.7. Next members discussed the requirements for each of the work holders and soldering blocks and had a volunteer draft the preliminary drawings. A final meeting was held with the vendor, after which the vendor produced final drawings, selected the material, and fabricated the respective fixtures.

	CAUSES		SNAGS AND DIFFICULTIES		RECOMMENDED ACTIONS
1.	Wrong orientation of wires inserted into connector	a)	Insertion of the 16 wires in their correct sequence into the connector wad done with the aid of the reference chart and by memory.	a)	Design a workholder for the connector with the appropriate colour of the 16 wires painted on it to act as guide for correct insertion of the wires.
2.	Wrong orientation of wires at soldering onto PCs.	a)	Plastic bags were used to store the 16 different colours.	a)	Design on appropriate workholder at soldering for the storage of the 16 wire types that arranges them in the appropriate sequence.
		b)	Soldering of the 16 wires in their correct sequence into the PCB was done with the aid of the reference chart and by memory.	b)	Modify the soldering block to include colour guides for the 16 wires.
3.	Improper fixing of wires into the connector (wires came out)	a)	Each crimped wire should be fully inserted into the connector until there is a "click" sound.	a)	Implement a "pull test" at wire insertion for every wire to ensure that each crimp is correctly seated in the respective sockets.
4.	Unsoldered leads.	a)	There is limited space on the PCB for tape adhesion when taping the display components for wave solder.	a)	Implement 100% inspection on all taped units before wave soldering to ensure that all the corner leads are not covered with tapes so that all leads will be soldered during wave soldering.

Figure 4.7 Determine corrective action.

Step 4a: Prepare instructions and flow charts for complicated procedures
Step 4b: Adequate training must be provided
Step 4c: Follow the plan exactly
Step 4d: Record any deviations from plan and collect data on results

- *QC tools that can be useful.* Checklist, checksheet, trend charts.

> **Example of Step 4, Implement Corrective Action**
>
> A training session was held for all production staff members to educate them on how to use the fabricated work holder and the soldering block. The recommended action was then implemented in work week 22 and results were tracked every day. The results were tracked on the graph shown in Fig. 4.8.

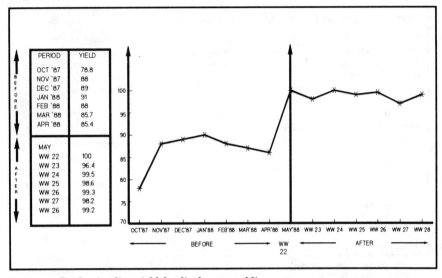

Figure 4.8 Production line yield for display assemblies.

Check stage

Step 5. Check the effect of corrective action

- *Objective of this step.* To check the effectiveness of the corrective action.
- *Discussion.* We now check the effect of the corrective action. There are several substeps that must be followed.

Step 5a: Compare overall result. Here we review the overall results. We must also review improvements on a paired Pareto diagram in order to compare before and after performance. Before and after results should be compared on all other items selected for study in step 2; use the same tools for making the comparison, such as bar graphs, paired Pareto diagram, trend charts, control charts, histograms, and process capability indices.

Step 5b: Failure to meet results. If failure is due to improper implementation, we must go back to step 4, implementation. Otherwise, we go back to step 3, analysis. If we fail to meet our goals, it is very likely that we missed the root causes, and further analysis will be required.

Step 5c: Results have been achieved; the goal has been met. If the overall result is equal to or better than the target set in step 1, we review the before and after data—especially the Pareto diagrams—and check that there are no side effects, that is, that there is no increase in the other categories of failures.

> **Example of Step 5, Check Effects of Corrective Action**
>
> A detailed analysis was done. The data was then plotted on a paired Pareto diagram. The results show a marked improvement in open digits. A check was made for unwanted side effects. There were none—in fact some of the other Pareto bars also decreased (see Fig. 4.9).

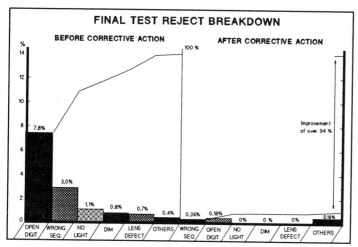

Figure 4.9 Paired Pareto diagram of before and after corrective action.

- *QC tools that can be useful.* Paired Pareto diagrams, trend charts, control charts, histograms, and process capability indices.

Act stage

Step 6. Take appropriate action

- *Objective of this step.* To ensure that the improved level of performance is maintained.

- *Discussion.* The corrective action that has been successful in improving performance must be documented in current operating procedures. There are several substeps to be taken.

    Step 6a: Documentation, standardize and control. The corrective action (implemented in step 4) that has been successful in improving the performance level should be documented in current operating procedures or standards. Poor documentation can result in problem recurrence in the future. It is very important to convey this information to other parts of the organization that may have generated the root cause of this

> **Example of Step 6, Take Appropriate Action**
>
> The team then recommended to the process engineer that the use of the respective fixture and the other corrective action steps be incorporated into the process specification. The results were conveyed to all employees and the importance of the new procedures were emphasized.

problem. Refer to the section "Standards Update Request Format" later in this chapter.

It is also important to identify the critical process parameters to control. The idea is to monitor the appropriate parameter and to detect any deviation from the new standards. Deviations should be promptly analyzed and eliminated.

Step 6b: Training. Ensure appropriate training in the new methods or standards. Employees must fully understand the changes that have taken place and the new procedures.

- *QC tools that can be useful.* Trend chart, control chart, checksheet.

Step 7. Decide on future plans

- *Objective of this step.* Use the experience gained for future projects.
- *Discussion.* The best area to look for a new project is the results in step 5. If our new trend chart or Pareto diagram has sharp peaks, we must eliminate them. If the bars in our new Pareto diagram have even heights, we must change our stratification base and then decide what to eliminate. If we have created side effects, we must work on eliminating them.

  We may also start afresh with a new breakthrough activity. But, first, we make sure that the entire process we have been through is documented according to the seven steps listed here. We make the recommendation because the documented project will be a good learning tool for new employees, and it will provide a historical record of improvements. The decision to continue with the current project or to select a new one has to be based on priorities and resources.

> **Example of Step 7, Future Plans**
>
> The team's future plans are as follows:
> 1. To continue monitoring of the first-pass yield of QDSP-6666
> 2. To move on to our third project
> 3. To achieve a 100 percent direct labor participation in QCC activities for our line (present status is 90 percent participation)
> 4. To extend the corrective action of this project to all cable-assembly-type devices in the production line

Seven Quality Control Tools and Other Methodologies

The seven quality control tools are relatively well known. They are as follows:

- Data collection, checksheets, and checklists
- Pareto diagrams
- Cause and effect diagrams
- Stratification
- Graphs and histograms
- Scatter diagrams
- Control charts

More details on each of these tools is given in App. 2. In Japan, a heavy emphasis is placed on education in and use of the seven tools. Often an analogy is made between these seven quality control tools and the seven tools used by the Samurai warriors of ancient Japan.

The Samurai warrior had seven tools, such as a sword, helmet, bow and arrow, and so on; he would never venture anywhere without these tools, which he needed for protection and success. In a similar vein, the seven quality control tools are essential for today's workers, engineers, professionals, and managers. They are needed for extracting information from data, conducting a proper analysis, and making correct decisions.

According to Kaoru Ishikawa, about 95 percent of the problems in the

> **Rate of Improvement and Setting Targets**
>
> What is a good rate of improvement? How aggressive should improvement targets be? For over 10 years we have come across several quality circles that use a rule of thumb of a 50 percent reduction in defects for each project, but we have never been able to trace the source of this rule.
>
> A recent article—"Setting Quality Goals" by Schneiderman in *Quality Progress,* April 1988—sheds some new light on this question. Empirical evidence suggests that most improvements (more specifically, reduction in defect levels) can be made at the rate of 50 percent in a very narrow time range of 6 to 9 months. Reduction of defects in an autonomous environment like manufacturing or administration takes a little less time. Reduction of defects in a complex environment such as between a factory and its supplier may take longer. The data shows that the rate of improvement is independent of cumulative volume—the learning curve—and is dependent only on time. The reason for this phenomenon is simple. Typically a few causes, the first few items on the Pareto diagram of defects, are responsible for a majority of the defects. Hence eliminating the first few items will quickly reduce defects by 50 percent or more. This cycle can be repeated several times.
>
> As an initial rule of thumb, we would recommend you plan to set a target of 50 percent reduction in defects every 6 to 9 months. By defects we mean waste, scrap, inventory levels, time to do something, and most other undesirable items. Once you set a target based on this simple rule, you should plan to deviate from it only if data suggests you do otherwise.

workplace can be solved using these tools. We have done an analysis at one Deming prize winning company—Yokogawa Hewlett-Packard—and are able to verify Ishikawa's statement. For the remaining, more difficult problems, you would use the seven new management tools, design of experiments, Taguchi methods, and so on. But, you would use them within the context of the PDCA cycle. Some details of the seven quality control tools and the seven new quality control tools are given in App. 2. For the others, the recommended readings in the Bibliography provide a guide.

Education of Employees

One of the important steps in the PDCA cycle is that of educating employees in the new improved process. This is crucial; otherwise the improvements cannot be maintained or the process controlled. It must be done quickly and efficiently every time.

We read with astonishment a remark made by the president, Boeing Aircraft:[5]

> After an unfortunate air crash, the error was traced to some workers' mistakes after a repair.... When the president of Boeing's Seattle plant was asked "How long will it take after re-education has begun before the technological strength [of your company] will begin bearing fruit?" His answer was seven years. Seven years! How can we ride around in jumbo jets for 7 years not knowing what types of defects they might have?

We hope this is an error. The comments here seem to reflect the different attitudes American and Japanese managers have toward problem solving. In one case a more leisurely pace seems possible; in the other, immediate attention and no recurrence seems to be the rule.

As we get continuous improvement, the process gets changed and redocumented and the work force needs reeducation. Because of growth or attrition, the work force needs continuous training—otherwise defects may increase. See the following box for a discussion on inadequate worker education and work standards.

Inadequate Worker Education and Work Standards

What happens when workers are poorly educated or trained or if work standards or procedures are inadequate? Figure 4.10 illustrates a real case of what happened when production volume picked up and new workers had to be hired. The new workers were inadequately trained or unaware of informal assembly procedures. The effect, in this case, was that all the results of continuous quality improvement were destroyed by inadequate work standards and training. Hence, what is required is "living" documentation—documentation that changes as the process changes and continuous training of workers and new workers. The same situation will occur in a service environment: Inadequate training will cause problems when customer transaction volume increases or employee turnover occurs.

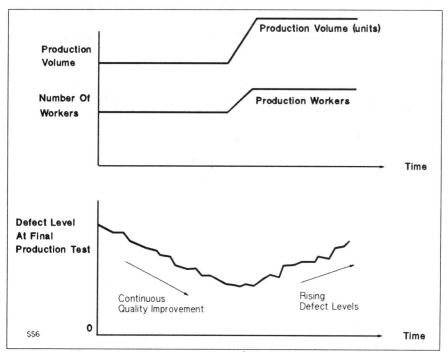

Figure 4.10 The effect of poor training and documentation on defect levels when production increases.

Source of Errors and Defects

What causes errors and defects? "The root cause is incomplete knowledge or imperfect work," says Hitoshi Kume. This causes problems everywhere, in every company and every country. Errors and defects will always occur. It is your job to minimize this occurrence. The introduction of standards will help reduce errors. This is the focus of the discussion in this chapter and the next, "Process Management."

Human errors

As you keep improving a process, each successive improvement gets more difficult. One of the fruitful areas to work on is the reduction of human errors. An effective technique to reduce human error is called *fool-proofing*—this ensures that smart machines which reject human misuse are employed. But you can't always resort to fool-proofing. You need to check causes of human error and determine the cause—lack of training, poor training, boredom, and so on. A good way to proceed is to use a human error troubleshooting tree to determine the source of the error. Figure 4.11, which was prepared by Katsu Yoshimoto of Yokogawa Hewlett-Packard,[6] provides some help. You will find it extremely useful as you attempt to get the last iota of improvement from a process.

One of the keys is setting up a fool-proof system. Many Japanese feel that the term *fool-proofing* could be offensive to some; hence Shigeo Shingo of Toyota Company coined the term *pokayoke* which means *mistake-proofing* or *fail-safing*.[7] Implementing pokayoke is an effective way to reduce human error and gets toward the goal of zero defects.

Where Can You Use the PDCA Cycle?

You can use the PDCA cycle for all problem-solving and improvement projects. For example, you can use it for improving your own work, improving product quality, improving service, improving marketing and sales activity, and reducing costs. The list is endless. Professor Iizuka of the University of Tokyo[8] has suggested four types of problem classification:

1. Reduction of existing defects
2. Improvement of a system, such as a quality assurance system or manufacturing process
3. Acquisition of new knowledge, such as optimal conditions in research and development or design improvement

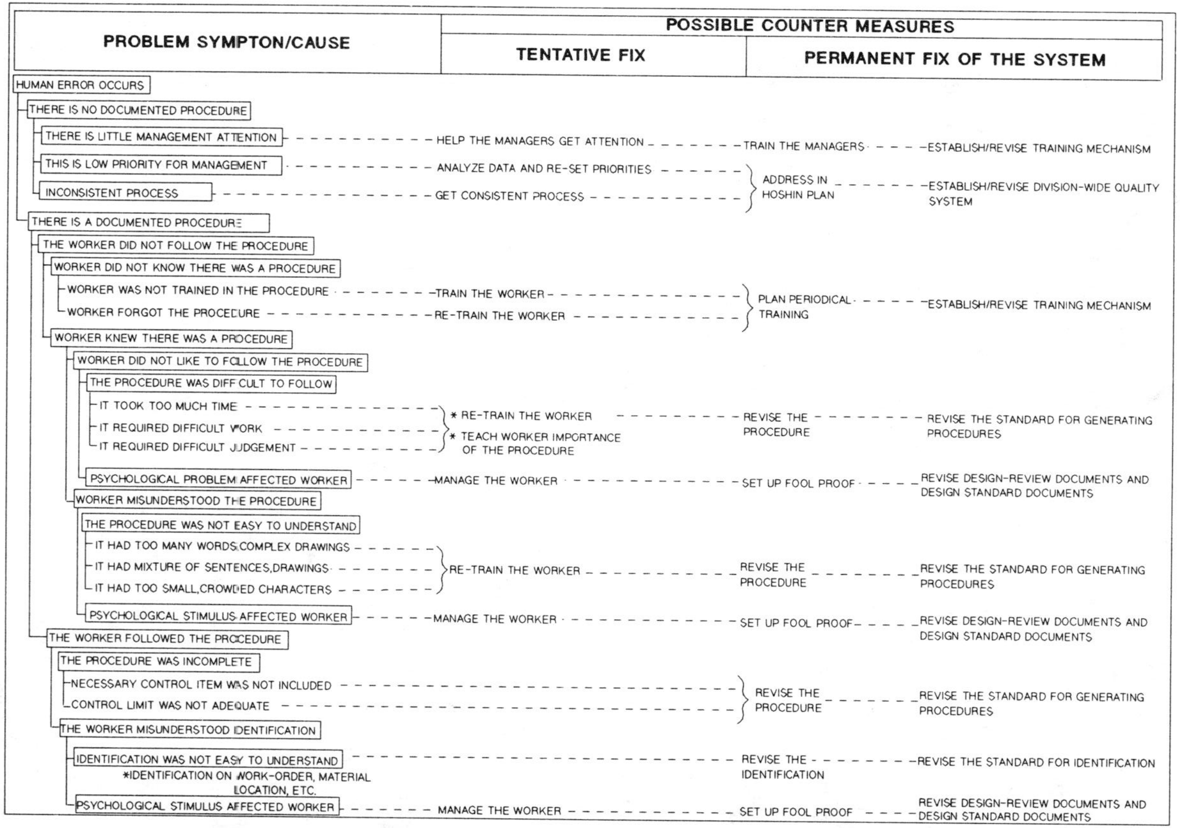

PROBLEM SYMPTON/CAUSE	POSSIBLE COUNTER MEASURES	
	TENTATIVE FIX	PERMANENT FIX OF THE SYSTEM

HUMAN ERROR OCCURS
- THERE IS NO DOCUMENTED PROCEDURE
  - THERE IS LITTLE MANAGEMENT ATTENTION — HELP THE MANAGERS GET ATTENTION — TRAIN THE MANAGERS — ESTABLISH/REVISE TRAINING MECHANISM
  - THIS IS LOW PRIORITY FOR MANAGEMENT — ANALYZE DATA AND RE-SET PRIORITIES — ADDRESS IN HOSHIN PLAN
  - INCONSISTENT PROCESS — GET CONSISTENT PROCESS — ESTABLISH/REVISE DIVISION-WIDE QUALITY SYSTEM
- THERE IS A DOCUMENTED PROCEDURE
  - THE WORKER DID NOT FOLLOW THE PROCEDURE
    - WORKER DID NOT KNOW THERE WAS A PROCEDURE
      - WORKER WAS NOT TRAINED IN THE PROCEDURE — TRAIN THE WORKER — PLAN PERIODICAL TRAINING
      - WORKER FORGOT THE PROCEDURE — RE-TRAIN THE WORKER — ESTABLISH/REVISE TRAINING MECHANISM
    - WORKER KNEW THERE WAS A PROCEDURE
      - WORKER DID NOT LIKE TO FOLLOW THE PROCEDURE
        - THE PROCEDURE WAS DIFFICULT TO FOLLOW
          - IT TOOK TOO MUCH TIME
          - IT REQUIRED DIFFICULT WORK } * RE-TRAIN THE WORKER / * TEACH WORKER IMPORTANCE OF THE PROCEDURE — REVISE THE PROCEDURE
          - IT REQUIRED DIFFICULT JUDGEMENT
        - PSYCHOLOGICAL PROBLEM AFFECTED WORKER — MANAGE THE WORKER — SET UP FOOL PROOF — REVISE DESIGN-REVIEW DOCUMENTS AND DESIGN STANDARD DOCUMENTS
      - WORKER MISUNDERSTOOD THE PROCEDURE
        - THE PROCEDURE WAS NOT EASY TO UNDERSTAND
          - IT HAD TOO MANY WORDS,COMPLEX DRAWINGS
          - IT HAD MIXTURE OF SENTENCES,DRAWINGS } RE-TRAIN THE WORKER — REVISE THE PROCEDURE — REVISE THE STANDARD FOR GENERATING PROCEDURES
          - IT HAD TOO SMALL,CROWDED CHARACTERS
        - PSYCHOLOGICAL STIMULUS AFFECTED WORKER — MANAGE THE WORKER — SET UP FOOL PROOF — REVISE DESIGN-REVIEW DOCUMENTS AND DESIGN STANDARD DOCUMENTS
  - THE WORKER FOLLOWED THE PROCEDURE
    - THE PROCEDURE WAS INCOMPLETE
      - NECESSARY CONTROL ITEM WAS NOT INCLUDED
      - CONTROL LIMIT WAS NOT ADEQUATE } — REVISE THE PROCEDURE — REVISE THE STANDARD FOR GENERATING PROCEDURES
    - THE WORKER MISUNDERSTOOD IDENTIFICATION
      - IDENTIFICATION WAS NOT EASY TO UNDERSTAND
        * IDENTIFICATION ON WORK-ORDER, MATERIAL LOCATION, ETC. — REVISE THE IDENTIFICATION — REVISE THE STANDARD FOR IDENTIFICATION
      - PSYCHOLOGICAL STIMULUS AFFECTED WORKER — MANAGE THE WORKER — SET UP FOOL PROOF — REVISE DESIGN-REVIEW DOCUMENTS AND DESIGN STANDARD DOCUMENTS

Figure 4.11 Problem solution model for human error.

4. Introduction or construction of a new system or drastic improvement of an existing system

To which we add a fifth item:

5. Managing an ongoing system, such as the Hoshin planning cycle. In the Hoshin planning cycle we need to plan, implement strategies and tactics, check progress, and take corrective action when problems occur. These are steps in the PDCA cycle. Look at Fig. 3.1; it is in PDCA format.

When you construct a new system (item 4 above), there is no need to have too precise an analysis. Since the goal is to have a new system or to drastically improve a current system, according to Yoshinori Iizuka, you should recognize existing deficiencies and agree on what you wish to achieve.

Let us elaborate a little more on the PDCA cycle for managing an ongoing system (item 5 above). In Fig. 3.3 we show the Hoshin objective to increase profits. That objective was deployed (in part) to the operation (or manufacturing) manager. In turn it was deployed to the operation manager's staff with the objective of reducing costs and failure rates. All that activity forms a vast PDCA cycle that needs good management, which we illustrate in Fig. 4.12.

On the left of the figure are listed the steps of the PDCA cycle as they apply to the objective, which was to increase profits. The operation manager adopted two strategies for achieving this: to decrease failures and reduce costs. Figure 4.12 shows how these strategies were deployed in the manufacturing operation and how results were monitored. The targets and the current status for each department are displayed on bar charts, and progress is measured regularly.

This is a modification of Komatsu's flag system,[9] which is used to manage their cost reduction programs. This is the detailed implementation plan of a Hoshin plan strategy, and it is a very powerful tool that gives impressive results. Komatsu has used this methodology, charts and all, as an effective management tool. The term *flag system* comes from the target versus actual bars which are displayed on a large chart or flag. These flags and corresponding plans are displayed and reviewed regularly. This helps to raise awareness of and commitment to achieving Komatsu's cost reduction program.

Example in PDCA format

During our discussions of the detailed PDCA cycle, we showed an example from a manufacturing environment. In App. 1, we provide an example from a sales environment—a follow-up to a customer satisfaction survey. Both examples follow the PDCA cycle steps, dis-

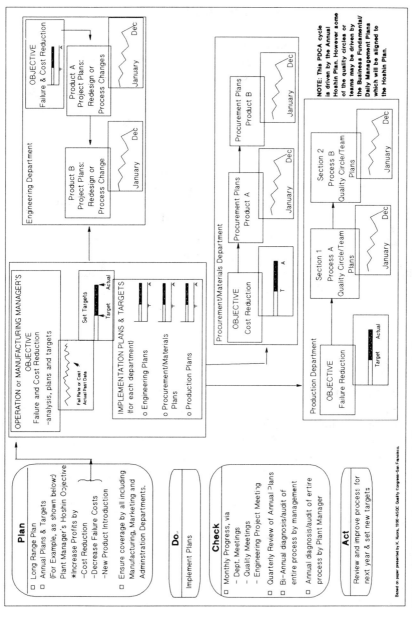

Figure 4.12 Using the PDCA cycle for management of an improvement plan.

cussed earlier. Study them and note how systematic the process is—how the current situation is analyzed and how the hypothesis is generated and then the hypothesis is validated. Only after study is a solution prepared. There is no fire fighting, no jumping to conclusions, no intuitive or genius approach.

We show in Fig. 4.13 the results of a major improvement project. This is a summary of the Yokogawa Hewlett-Packard (YHP) solder wave project.[10] This project has become a legend in the quality community, and it has been shown countless times within and without the Hewlett-Packard Company. There are two interesting points to this project, cited by Kenzo Sasaoka, President of Yokogawa Hewlett-Packard:

- Because of the success, "the soldering process became the catalyst of the TQC movement"[10] at YHP.
- The soldering machine used was not state of the art but instead "a jalopy" transferred from another plant more than a decade before, "which never fails to astound every" visitor.[10]

Notice that YHP went through three major improvement cycles and was not satisfied until it reached a failure rate of 4 parts per million. That is equivalent to finding 4000 specific people in China's population of 1 billion. Think about that. More important, when would you have been satisfied? After the first or second cycle or would you have gone on and on like YHP?

Standards

We discussed the need to have a standard improvement methodology—the PDCA cycle. Everybody in the organization should use the same improvement methodology. This concept is similar to one of having standards in all areas of commerce and industry—standard electrical sockets, standard material for building a house, and so on. This makes shopping, planning, and work so much easier. Similarly, we need standards in the TQC environment.

Standards represent proven best practices that are institutionalized in an organization. All employees must be trained to understand and use the standards that are relevant to them. Good standards will save time since employees will not need to reinvent the wheel in other parts of the organization. Instead they can use their creative abilities to invent things in new, unexplored areas. This will result in optimum use of your most valuable resource—all your employees.

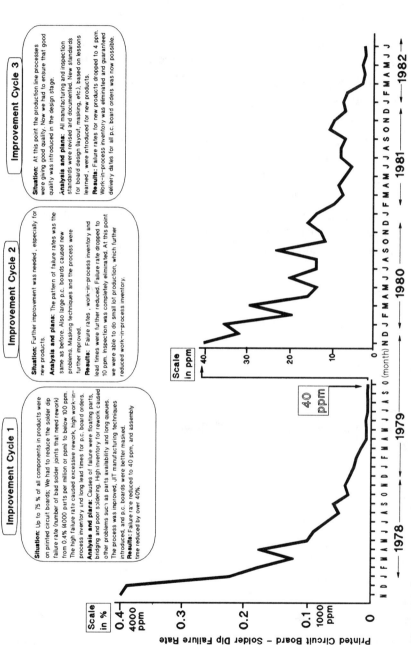

Figure 4.13 Improvement of printed circuit production process.

Benefits of standards

Let us restate the benefits of standards or proven best practices:

- Standards minimize the need for new resources and ideas—just use a standard form or a standard procedure across an entire organization.
- Standards minimize human errors and other types of defects, thus saving time and reducing rework.
- Standards make it easier to communicate ideas and information.
- Standards when established allow time for creativity in other new areas; otherwise time is spent in:
  - Slower problem resolution, if the standardized PDCA improvement cycle is not used
  - Redesigning where there are no design standards, for example, component or reliability standards
  - Retraining where training is poor and not documented, for example, operator or designer training
  - Longer time to learn when managers or employees move to different locations in an organization using different standards for planning, problem solving, designing, and so on

Important points for creating standards

The following pointers will be useful when standards are created and maintained:

- Create standards which can be followed.
- Revise standards which are difficult to follow.
- When training workers to follow standards or procedures, consider their receptivity and ability to follow them.
- Observe if standards are being followed, and understand why workers do not follow them.
- The standards must be easily accessible by designers, engineers, buyers, and other company employees. More on this when we discuss standards update requests later in this section.

Preventing recurrence

One of the steps in the PDCA cycle is to standardize, control, and document the new improved process. How important is this? Extremely. Let us illustrate the importance with a true story:

Recently, I was reviewing a manufacturing operation by conducting a TQC review. One of the improvement projects they showed me was entitled "Reduction of Scratches in Display Modules." They manufactured and shipped display modules for instruments and aircraft cockpit displays. These got scratched during packaging, when they were inserted in shipping tubes. As a result of improvements they were able to reduce scratched displays to zero. Very impressive.

But wait a moment. I have a long memory; and I remarked that many years ago—maybe 6—I had reviewed a similar project for a different product. Then, as now, there was a project to reduce scratched displays, caused by shipping tubes. Why had they not learned from their errors? Why had they not conveyed the lessons learned to the designers of the shipping tubes? I told them: "If you don't learn from your errors, if you don't convey the lessons learned into new standards, you are doomed forever to have projects to reduce scratches in display modules."

Does this sound familiar? Every company has a similar situation. The outcome is problems that keep recurring throughout the company and improvement projects that are repeated, resulting in lower productivity and customer dissatisfaction.

The solution is to convey the lessons learned into standards or to upgrade existing standards. How do you upgrade existing standards? We propose the use of a standards update request.

Standard update request format. Often, there will be a need to update standards—especially design standards. This must be done via a formal process rather than a verbal request or complaint. In the case of the customer complaint and feedback system, we suggested the use of a corrective action request (CAR) to formally drive the change in the organization. Here we propose the use of a standards update request (SUR) format that can be used to drive the change. The format used can be on paper (in the beginning) or via an electronically transmitted medium. We show a recommended and completed format in Fig. 4.14.

In the figure, the proposal to update the standard is conveyed to the standards owner in the company. This could be a R&D standards manager, a specification control manager in the purchasing or materials department, and so on. The manager must ensure that the standard is updated by modifying design software or changing drawings and specifications. In all cases the new standard must be documented and filed—preferably in a computer file. The use of a keyword—as recommended in the SUR format—will make it easy for designers, buyers, and other company employees to review design standards.

Sources for standard updates are as follows:

The improvement cycle. When an improvement project is completed and if an improvement in design is proposed.

Customer complaints and feedback. When customer complaints are received, problems are analyzed, and an improvement in design is made. We show an example of this in the completed SUR format in Fig. 4.14.

Daily Process Management. As processes are managed, problems and errors may be discovered. For example, during a project postmortem, weaknesses in the original design may be discussed and a SUR generated.

This concept of standards will be revisited in the next chapter, "Daily Process Management." There we will focus on standard processes: processes that are well understood and documented, processes that stay even as managers and employees come and go.

Problem-Solving Hierarchy

As we have mentioned, problems and errors will always occur. Our goal, then, is to minimize problem and error occurrence. As we decrease our errors and eliminate problems, we will go through the following phases or hierarchy:

1. *A fire-fighting mode.* Here problems occur helter-skelter throughout the organization, and we fix them on an ad hoc, nonsystematic basis. Everything is disorganized. Often, the bearer of bad news is punished. The person who fixes the problem and douses the fire is rewarded and considered a hero; but no attempt is made to understand why the fire occurred or why there are so many fires.

2. *Systematic problem-solving mode.* Here we are in control, fixing problems systematically, eliminating root cause, and operating in an environment of continuous improvement. The PDCA cycle is used constantly and consistently; systems are in place to detect problems and resolve them quickly.

3. *Prevention of problems by prediction mode.* Here we go beyond systematic problem solving; instead we predict and prevent potential problems from occurring. We are able to understand customer needs before we design a product or service—this will prevent low customer acceptance and subsequent design changes. Also, before the release of a product or service, we are able to predict potential prob-

STANDARDS UPDATE REQUEST (SUR)	
STANDARDS OWNER: R&D department	MAIL TO: R&D Section Manager
Item Printed circuit eyelet	
Product: Display Module Originator Product Engineer	Category _____ Date January 10, 1991

KEY WORDS: (FOR COMPUTER FILE)
    Eyelet, through-hole

SUMMARY OF ISSUE: (ADD MORE DETAILS, WHERE APPROPRIATE)
Our customers have reported several cases of open circuits on Display modules that were supplied by us. Printed circuit board traces were found to be open near the eyelets or through-holes, after the customer had inserted pins on the display module. The analysis shows that the printed circuit pad is too small and the neck breaks during insertion. Failures are estimated at 100 parts per million and can be reduced to zero. (Analysis and photos are attached).

PROPOSAL: (ADD DETAILED PROPOSAL, WHERE APPROPRIATE)
Change the shape of the printed circuit eyelet pad as shown. This will increase pad area by about XX % and adhesion by YY %. This is expected to reduce defects to zero ppm. Detailed calculations are attached.
Current pad shape   neck   Proposed pad shape

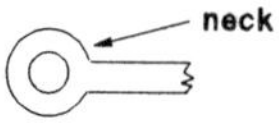

COMMENTS BY STANDARDS OWNER:
Analysis is appropriate. We agree to change all through-hole / eyelets for pin insertion. This change will be valid for all designs. Software will be modified to change current design rules.

Approval:
Section Mgr: _____
Dept Mgr: _____
Action By: _____
Date: _____

Figure 4.14 A standards update request format.

lems and eliminate them. Some tools that will allow us to do this are quality function deployment (QFD), and failure mode and effects analysis (FMEA). These and other tools are discussed in Chap. 5.

Clearly, this last phase is the one we want to be in—always.

Questions and Answers on the Improvement Cycle

Let's review some frequently asked questions on the improvement cycle.

Q1. Why should we restrict ourselves to one specific improvement methodology (that is, PDCA)? This will restrict our creativity.

A1. If there is a rigorous, proven, and successful improvement methodology, you should adopt it instead of inventing your own. There are two advantages to doing this. If there is a standard methodology in your company, and everybody trained to use it, you have a common problem-solving language in your company. There is no need for retraining as employees move across the company. Creativity is very important. But it should be used in new unexplored areas or in providing creative solutions that thrill your customers. Hence you will not stifle creativity; instead you are pointing it in the right direction.

Q2. How do we ensure that we improve the right things?

A2. Priorities need to be set. In the beginning of this chapter, we discussed how you can select items to improve. This will form your initial list. From that you develop priorities, which are influenced by the following:

- Linkage to the annual Hoshin plan. That is, is this an item that shows up in or supports a Hoshin plan strategy?

- Customer complaints. Frequent customer complaints must be given top priority.

Q3. Must we document every completed project in the PDCA format?

A3. No. The important point is that for every project the PDCA framework is followed. Completed projects should be filed and sorted in PDCA categories. That is, you can access the project targets, schedule, analysis, implementation plan, check data, and so on. However, one or two projects should be documented in the PDCA format, examples of which are in App. 1. These documented projects can be used for training, promotion, publicity, and presentations.

Q4. You quoted Kaoru Ishikawa as saying that the PDCA cycle is the most important thing in TQC. Does that mean that improvements are the most important part of TQC? If so, what about new product design and planning for the future—aren't they important?

A4. They are very important. Continuous improvement is a basic tenet of TQC; and the PDCA cycle plays an important part. But the PDCA cycle can be used for managing other activities as well: the long-term plan, the annual Hoshin plan, projects in R&D, and education programs. The PDCA cycle can also be viewed from a macro angle. For example, for a design and manufacturing company, P is the design stage, D is the manufacturing stage, C is the selling and getting customer acceptance stage, and A is the customer feedback stage. That brings us back to P, or the next design.

Summary: The Improvement Cycle

We have discussed and emphasized the importance of a systematic improvement cycle or process. We recommend the use of the PDCA improvement cycle for all improvement projects. This cycle is especially useful because it can be used to manage each improvement cycle as a project with schedules and targets. In addition it helps to identify root causes of problems and removes them. Within this cycle you can use the numerous statistical tools such as the seven tools, the seven new tools, design of experiments, and Taguchi methods.

It is important to adopt the PDCA cycle as a standard process for all improvement projects. Otherwise you may get less than optimum results, resulting in constant fire fighting. You should really discourage the use of other improvement frameworks or templates. There is no need to be creative in designing an improvement framework. Creativity can and should be encouraged in other unknown areas where there is no experience or knowledge—for example, in providing creative solutions to your problems or alternatives for your customers.

We have discussed the importance of standards. In addition to adopting the PDCA cycle as a standard, there should be standards for all repetitive activity. We have suggested a formal procedure to update standards in the entire organization—we proposed the use of a standards update request. Standards are very important, and we will discuss them in more detail in the next chapter.

Finally, we discussed a problem-solving hierarchy, which progresses from the fire-fighting mode to systematic problem solving with the PDCA cycle to prevention of problems by prediction. Our ultimate goal is to predict and prevent problems from occurring. This is the phase we want to be in—always.

References

1. Juran, Joseph, Frank M. Gyrna, and R. S. Bingam. *Quality Control Handbook.* New York: McGraw-Hill, 1951.
2. Feigenbaum, A. V. *Total Quality Control.* Singapore: McGraw-Hill, 1986.
3. Kume, Hitoshi. "Business Management and Quality Cost, The Japanese View." *Quality Progress,* April 1985.
4. Deming, W. Edward. *Out of the Crisis.* Cambridge, MA: MIT Press, 1982.
5. Comments taken from "The Japan That Can Say No" by Akio Morita and Shintaro Ishihara. Anonymous English translation.
6. Prepared by Katsu Yoshimoto of Yokogawa Hewlett-Packard. Adapted from "Principles of New Product Development," by Iizuka, Yoshinori. ICQC, Tokyo, 1987.
7. Nikkan Kogyo Shimbun Ltd (eds.). *Poka-Yoke: Improving Product Quality by Preventing Defects.* Cambridge, MA: Productivity Press, 1988.
8. Iizuka, Yoshinori. "Key Points for Success in Problem Solving." *1990 ASQC Quality Congress Transaction,* San Francisco.
9. Kuroa, Kozo. "Survey and Research in Japan Concerning Policy Management." *1990 ASQC Quality Congress,* San Francisco.
10. Extracted from an article by Kenzo Sasaoka, Yokogawa Hewlett-Packard. "A Challenge to Revitalize Management." Quality month text no. 155, published by JUSE in Japanese.

Chapter 5

Daily Process Management

People come and go but processes stay.
　　　　　　　ANONYMOUS

Overview

In Chap. 3, we discussed Daily Management, or Business Fundamentals, plans. The Daily Management plan lists items or key processes that are managed on a day to day basis. Progress is measured by selecting appropriate performance measures, setting goals, and reviewing them regularly—often daily. In this chapter we will review in detail how some of the key processes in the design, manufacturing, and sales environments can be managed or controlled.

In Chap. 4, we discussed the relationship between improvement and control. All the key processes in an organization must be controlled. The following incident illustrates the point.

> My friend and mentor Noriaki Kano, a member of the Deming prize committee in Japan, was once reviewing a manufacturing operation. He had been impressed by everything the managers had shown him. As they passed a production operator working on a bonding machine, he stopped to talk to her.
>
> He asked her if she knew the reject rate of her bonding machine. "Oh yes, she replied, it is 0.15 percent." What did she do if for any reason the rejects exceeded that number? She replied that she had been told to call the supervisor whenever rejects exceeded 0.20 percent. Kano decided to check the records to verify her statement. He flipped through her records and found a day when the lot reject percentage was 0.24 percent—so he asked her, "What did you do?" She replied that 0.24 was so close to 0.20 that she did nothing. "Fair enough," he said and looked at the records again and found another lot about 1 week before with a reject rate of 0.30 percent. "What did you do then?" he asked. She replied that she had indeed informed her supervisor.

The supervisor happened to be standing by and informed us that she had told the engineer. Kano remarked, "It's been a week, what has the engineer done?" The engineer was nowhere to be seen. He suggested looking for the engineer and sent someone to find him. Meantime, the production manager was getting a little nervous and suggested they move on. Kano stood his ground and kept asking to meet the engineer. We finally found the engineer. He had been informed but what had he done? He had done nothing, no documentation or analysis was available; he just never got around to doing anything, but he planned to do something—when he had time.

The learning points of this episode are:

1. A good job of characterizing the process had been done per the textbooks.
2. When the process went out of control, the company never had the discipline to analyze and take corrective action.
3. Eventually the process or the bonding machine could have given more rejects or gone out of control, but nobody would have noticed—until it was too late.

After this experience, Kano commented, "The workers are awake, the engineers are just waking up, but management is asleep." His point was that the workers were doing their best and alerting management to potential problems, the engineers were busy and stretched, and management was oblivious of the real problems. In our example, there was a possibility that if the current neglect continued, problems would multiply and the process would run out of control. The outcome could have been product problems and customer dissatisfaction. Here is where we can apply the concept of daily process management.

Daily Process Management

Daily process management means identifying and monitoring a process, ensuring it meets a target (within limits), discovering abnormalities, and preventing their recurrence.

The process concept

Before we discuss daily process management further, let us discuss the process concept. Every activity is part of a process. This applies to the manufacturing and service sectors—in fact, any activity at work or home. Figure 5.1 shows a simple process flow.

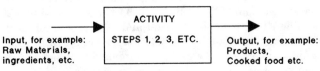

Figure 5.1 A simple process flow.

To improve the quality of a product or service, the process must be improved by changing or modifying it. We discussed improvements in the last chapter and showed how improvements and control are linked. We mentioned that every process must be controlled or managed. In part, we manage a process in order to retain the gains after improvement; but we also manage a process to better control its output or results.

Why manage processes? Achieving good results is paramount in any company—because they indicate commendable performance, which is what management strives to have. Good results, however, are a *lagging indicator* of performance; only when results are achieved do we know that we have performed well. But we must be able to predict results—for this we need a *leading indicator*. A well-managed process, monitored with properly selected performance measures, can be a leading indicator and will give predictable results. For example, if a prize-winning recipe is followed diligently, it will give an outstanding dish. Similarly, good design, manufacturing, and sales processes will give predictable and outstanding results in any company.

To reiterate, a well-managed process can be a leading indicator of good results in any company. In this chapter we will discuss how to control or manage a process. Given that every activity is a process, we will identify key or important processes in the design, manufacturing, and sales environment that must be managed. We will also discuss how to manage processes on a day to day basis, hence our use of the term *daily process management*.

The requirements for daily process management

What are the requirements of a good process? For a start, the process should be documented, everybody should be trained, and the performance measures to monitor performance of the process should be well understood. If the process goes out of control, corrective action should be taken to bring it back into control.

Daily process management at a factory. Let us look at an example at a facsimile machine factory. In it there could be a printed circuit board assembly line and at the end of the line a test station. One of the performance measures for the process could be failure rate at the test station. Figure 5.2 shows what to expect with an unmanaged process, a well-managed process, and an improved, well-managed process. The lower the failure rate, the better the quality of the printed circuit boards coming off the line. The result is less rework, higher productivity, and a predictable output from the production line.

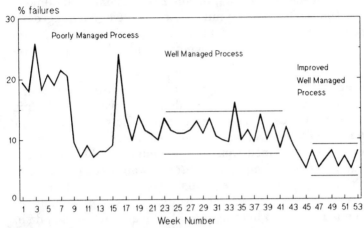

Figure 5.2 Measuring and managing failure rate.

Daily process management in a marketing and sales operation. The same concept applies to the sales process in a marketing and sales operation, as shown in Fig. 5.3. Assume one of the performance measures in the operation is the success rate, defined as

$$\text{Success rate} = \frac{\text{number of qualified prospects who purchase}}{\text{number of qualified prospects}}$$

Doing well here means that the marketing and sales people have done a good job of selecting, screening, and qualifying prospects during the preselling or marketing stage. The result is that the chances of selling are now higher and more predictable. Less energy is expanded and productivity is higher.

Our purpose is to aim for a well-managed and predictable process by documenting, monitoring, and controlling the process. When any key process becomes stable and well managed, it becomes part of the Daily Management, or Business Fundamentals, plan as an entity. Managers can then use their creative energy for other problem processes or new unexplored areas. See the comments given in Chap. 4, "The Improvement Cycle" about standardization.

Prerequisites for good process management

How do you ensure that a process is well managed? What are some of the necessary prerequisites. Here is a short list that is a good guide:

1. *The process must be documented and must include a process flow chart.* The documentation includes key steps with performance measures for each step.

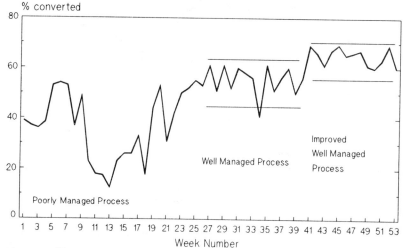

Figure 5.3 Measuring and managing success rate.

2. *There must be process performance measures, with targets.* Refer to Fig. 5.1. Each of the steps (1, 2, and 3) could have a target which must be consistently reached. So, for example, in the facsimile machine factory which we discussed earlier, step 4 could be the test station. The performance measure could be board failure rate—with a specific target of, say, 2 percent. If the other steps are well defined and understood and managed like step 4, we have a well-managed and predictable process in the factory.

3. *There must be a method to manage out-of-control conditions.* In managing a process, the concept of statistical quality control must be understood and applied. We discussed the concept of limits in Chap. 3, "The Planning Process." If you are still unfamiliar with this concept, refer to App. 2 and the Bibliography.

Let us go back to Figs. 5.2 and 5.3. In the left of both figures we have a very unpredictable, wildly gyrating or poorly managed process. Using the concept of statistical quality control to eliminate assignable causes (such as poor training or inferior machines) will result in a well-managed process. We can see from the charts that the process is fluctuating between two statistically identifiable limits—the upper and lower control limit. When the process is contained within these limits, we have a managed process. We can further improve the process by understanding the reason for failure—for example, need better raw material for the printed circuit boards or need a better way of qualifying prospects, perhaps by understanding their needs better. If we are able to do this, we will get an improved well-managed process as shown on the right sides of Figs. 5.2 and 5.3.

We will find that occasionally the process will deviate and cross the control limits. When this occurs, it is important to analyze the reason for the deviation and to take corrective action to prevent recurrence. Otherwise this may be the start of a deteriorating cycle. Failure to understand and correct a deviation can also give the wrong signal to employees—that is, if there is a deviation, we can ignore it. There will be more on deviation analysis later in this chapter when we discuss out-of-control reports.

4. *There must be a training plan.* All employees must be trained in the appropriate process steps that are applicable to them. And if the process is changed or improved, the changes must be documented and employees trained in the improved process. This is one of the steps in the PDCA cycle.

5. *The process must be continuously improved.* An aggressive management will continuously set new and tougher targets for each important performance measure in the process. Therefore, not only must a process be managed, but it must be continuously improved. Here lies the secret of success for any company.

Identifying key processes to be managed

In any company there are key processes beginning with customer needs and ending by meeting those needs. For example, in research and development (R&D) the product is designed based on initial customer needs; it is then manufactured and finally delivered to the customer.

Let's illustrate and explore this concept further. In Fig. 5.4, we show the key processes in a design, manufacturing, and sales environment. There are others that we have left out, but we wish to focus on the vital

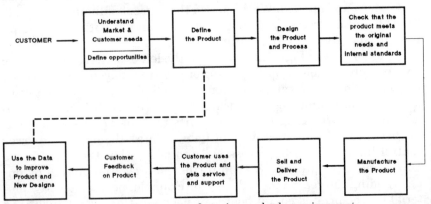

Figure 5.4 Key phases in a design, manufacturing, and sales environment.

few—the few that have the most impact. If we agree that these are important and necessary, we need to understand the processes or phases from product conception to product delivery. These processes must be documented, well managed with performance measures, and improved. Progressive improvements will provide faster throughput, resulting in shorter time to market.

This is the way to get a more nimble organization that is less dependent on organizational change or people and management turnover. By this we refer to one of the greatest problems of organizational change and people turnover: A new manager or employee will bring along his or her "pet" process or idea and change the basic process that exists. This constant turmoil is certainly very destructive and leads to a less than optimally managed organization. Figure 5.5a and 5.5b shows specifics of the concepts explained in Fig. 5.4. Some of the key processes in the design, manufacturing, and sales environment are shown. In the next few sections, we will discuss most of them, and for several we will go into detail and suggest how you should manage them.

Competitive Benchmarking

We discussed competitive benchmarking in Chap. 2, "Customer Obsession." This is an important process, because we must be aware of what world-class companies and competitors are doing—new ideas, material processes, and manufacturing techniques that are being deployed. Some of the data obtained will be charted for comparison during the product definition process of a new product or service.

The Product Development Process

The product definition process (which is the beginning of the product development process), if done well, will ensure that the right products are manufactured. The succeeding process (design) will ensure that the products are designed correctly—with appropriate features and high reliability.

One of the best ways of ensuring a good product definition process is by using the Quality Function Deployment (QFD) methodology. QFD is a tool that helps to tightly link the entire product development process. It enhances the product definition process and also ensures that the succeeding processes, such as design and manufacturing, are linked to product definition. We will discuss QFD in the context of a good product development process. The discussion on QFD comes courtesy of Khushroo Banu Shaihk of Hewlett-Packard.

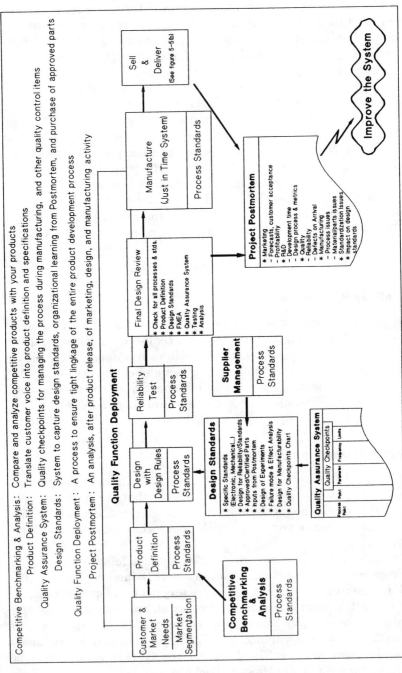

Figure 5.5a Key processes in a product design and manufacturing environment. These represent the vital few and are not in sequence.

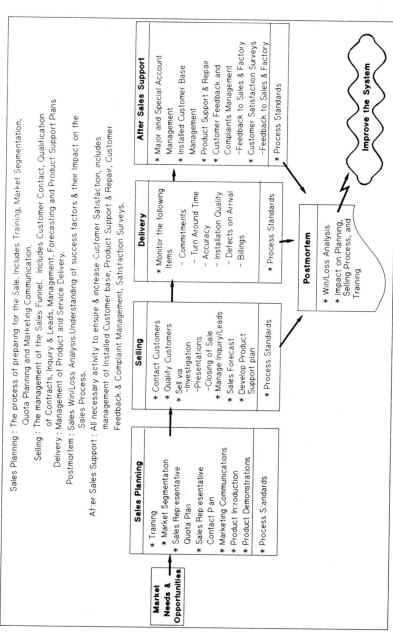

Figure 5.5b Key processes in marketing, sales and after-sales support. This is a continuation of Fig. 5.5a and represents the vital few.

Introduction to QFD

Why QFD? A lack of a well-defined product development process in R&D can be the root cause for the ruination of a company. QFD offers many features which are helpful in defining this process, something that for many reasons is often very difficult.

Although many companies do not have a well-defined new product development process, they manage to market blockbuster products or services which fill the needs of customers or users. At other times their new products or service may fail because customers do not purchase them. In other words, success comes when customer needs are well understood and met by the products or services. This is another powerful feature of QFD. Hence, learning more about QFD is worthwhile. We mentioned earlier, in Chap. 2, the reasons why products or services succeed or fail in the marketplace. A common thread runs through all the studies that we cited: *To have a successful product, we must understand user needs.* QFD provides a formal process to understand user needs and to weave these needs across the entire product development cycle, including manufacturing.

Definition of QFD. QFD is a very systematic and organized approach of taking customer needs and demands into consideration when designing new products and services or when improving existing products and services. Another name for this approach is *customer-driven* engineering because the voice of the customer is diffused throughout the product (or service) development life cycle.

QFD is a planning tool that defines a process for developing products or services. The aptitude to plan is rare in the human race. Managers are evaluated on short-term results, which further inhibits this aptitude for planning. It is difficult to use Deming's PDCA cycle to improve the product development process if the P of PDCA cycle is weak. QFD is applying TQC philosophy to product development by focusing on P. Using QFD counteracts the inherent weakness embedded in human nature—that of avoiding planning.

Where did QFD come from? Yoji Akao of Tamagawa University is the key contributor to QFD development in Japan. There are 30 matrices in his approach. The QFD team can pick and choose the matrix which would be of most use for a particular phase of product development. There are many other experts; for more information on the history of QFD you should read a chapter written by Akao.[1]

Another expert in the area of QFD is Fukahara. He is associated with the Central Japan Quality Control Association. He mainly focuses on the house of quality, namely the product definition aspect. There is also the four-phase (or four-matrices) approach promoted by the American Supplier Institute (ASI). Later we will discuss both the house of quality and the four-phase approach. There are, of course, many other approaches to QFD.

Each approach has its pros and cons. What is the best approach? The answer is not very straightforward. Each company has a unique set of employees, products, customers, systems, and so on, giving birth to unique circumstances and culture. The best solution is to develop a custom-fit QFD model to meet the particular needs of your company. If a cookbook approach to apply a tool like QFD is taken, your company will not get the optimum benefit of QFD but instead may get bogged down.

The collective brain power of the employees is the most valuable resource of any company. One of the major goals of using TQC philosophy is to enable employees to make meaningful contributions as a result of using their collective brain power. QFD allows this freedom.

QFD in the United States. The principles and ideas used in QFD were suggested by A. V. Feigenbaum when he discussed total quality systems and a cross-functional management way of thinking. Joseph Juran describes quality assurance systems in his *Quality Handbook* and touches on the early stages of a QFD system. Naddler, a systems dean at University of Southern California, teaches "The Planning and Design Approach" with ideas that could have culminated into QFD as it is today.

The first formal introduction of QFD as whole system came to the United States in 1983, via the Cambridge Corporation in Chicago. The following year Don Clausing from MIT introduced QFD to Ford Motor Company. Don Clausing and John Hauser wrote an article about it in the *Harvard Business Review* in May of 1988. In the same year Ford told its vendors that in order to remain on Ford's preferred vendors list, it would be advisable for them to use QFD. Symposiums on QFD are regularly held these days. QFD is spreading like wildfire in the United States.

What are the benefits of using QFD? The ultimate benefit of QFD for any company is its contribution to meeting and exceeding customer needs, thus obtaining higher market share and profits. There are, however, many other tangible benefits that participants can experience. Some of them are listed here:

1. Product definition is firmer and takes place earlier in the new product development life cycle. This minimizes engineering changes and results in better quality.

2. QFD addresses major issues and complaints expressed by customers during the early stages of product definition. Hence, the number of complaints about and dissatisfaction with new products decreases with time. This benefit is seen after several product cycles.

3. Cross-functional walls break down with QFD since the team must address issues that affect all departments. Suboptimization of resources in a company is minimized and communication between departments improves.
4. Team members develop a deeper understanding of customer needs and have the customer's voice as a basis for making tradeoffs, resulting in superior decisions for the organization.
5. The analytic vigor of QFD causes streamlining or elimination of many internal processes that do not add value to the new product development process.
6. Customer needs are evaluated with respect to competitive products and services. This allows identification of the internal processes that need improvement to stay competitive.
7. Documentation is an essential ingredient of QFD. Hence, one of its greatest benefits is that we build product intelligence. This documentation provides the following advantages:
   - It helps new engineers come aboard faster.
   - Easily accessible documentation reduces chances of repeating mistakes of the past.
   - The accumulation of knowledge decreases the need of having someone with seniority lead the project. A senior leader contributes significantly to the success of the project, as we learned from the project Sappho study in Chap. 2. A well-established QFD process will deliver this critical benefit—even from a younger person who possesses leadership qualities. Also, under a TQC environment the team will feel empowered and possess the authority generally associated with having a senior leader.
8. QFD is beneficial in understanding and identifying a market niche where customer needs are not being met. This provides opportunities for introducing niche products.
9. QFD provides an excellent framework for cross-functional deployment of quality, cost, and delivery. The importance of this was discussed in Chap. 3.
10. QFD allows for quick changes, which is very important for the new product development process. It is possible to revise previous decisions when new information becomes available during product development, for example, if the competition introduces a new product or a state of the art technology becomes available. Detailed documentation keeps all information visible to the QFD team at all times.

All of the above benefits result in a robust new product development cycle and minimize difficulties and problems. QFD is one of the best ways to introduce TQC to the marketing, design, and manufacturing environments.

Implementing QFD is a very strong strategy for propagating TQC in a company. QFD provides the much needed horizontal weave across the company. Typically, this horizontal weave is missing in the current traditional management hierarchy because organizations are managed vertically. The benefits of QFD will include a shorter product or service development cycle and higher productivity.

Nuts and bolts: How to begin. The ASI, the Kaizen Institute, and the Greater Opportunity Alliance of Lawrence (GOAL) teach the nuts and bolts of QFD in the United States. The details of how to implement QFD can be learned from attending their classes. Some books have been published on the subject,[2] but we recommend you attend a QFD class to acquire hands-on knowledge of this subject. We will share a brief overview of the process, using the example of the pencil. We will also share insights developed from working with QFD and in some cases bring more depth to a particular topic. This will enhance your chances of developing a dynamite product or service.

The four phases of the product development cycle are listed below and shown in Fig. 5.6:

I. Product planning
II. Part deployment
III. Process planning
IV. Production planning

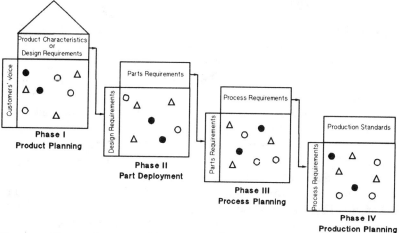

Figure 5.6 Phases of customer-driven product development.

Figure 5.6 illustrates the closely linked phases of QFD as the product develops. The tightly linked characteristic of phases accomplishes delivery of the product as desired by the customer. The design requirements of the first matrix feeds into the second matrix so that part requirements are done correctly. Part requirements then drive the process requirements of the third matrix. Process requirements for manufacturing will determine process assembly and the quality checkpoints chart. Details of quality checkpoints are given later in this chapter and will ensure consistent product quality.

The critical planning for the product takes place in the first phase. This phase is also known as *the house of quality matrix* and *matrix of matrices*. The components of this matrix are:

1. Customer's voice
2. Importance rating
3. Product characteristics or design requirements
4. Relationship matrix
5. Quality planning using competitive evaluation and strategic positioning
6. Correlation matrix (the triangular roof portion)
7. Technical evaluation
8. Target value determinations

A house of quality with these components is shown in Fig. 5.7. The house of quality, which focuses on product definition, is a critical engine that guides all steps of product development. The overall success of the product therefore depends on good product definition.

Why is product definition important? Research indicates that schedule slippage in product development occurs because of either lack of firm definition or changing definition of the product. So product definition is a critical first step in making a good product. QFD models of Fukahara, Makabe, and Yoji Akao start with a matrix that facilitates product definition. Freezing the definition helps the development of the product in a timely manner. Additions and deletions not only cause schedule slippage but have a deteriorating effect on quality. One possible outcome of product definition is a decision not to make the product. This decision reduces the cost of abandoning projects later. Abandoned projects can be a tremendous drain on business, as noted in the Union Carbide study that is discussed in Chap. 2.

HOUSE OF QUALITY

```
                    Correlation Matrix
                  ┌──────────────────┐
                  │Product Characteristics│  Competitors
                  │                  │   X  Y  Z
     ┌────────┐   │ I              │  ┌──────────┐   ┌──────────┐
     │--------│   │ M    O         │  │          │   │ Strategic│
     │--------│   │ P  △ Relationship│ │Competitive│──│Positioning│
     │Customers'│ │ O    △         │  │ Evaluation│  └──────────┘
     │ Voice  │   │ R  ● Matrix    │  │          │   ┌──────────┐
     │(Needs &│   │ T  △ △  O      │  │          │\  │Merchandising│
     │Demands)│   │ A    △         │  │          │  \│ Strategy │
     └────────┘   │ N              │  └──────────┘   └──────────┘
                  │ C              │
       Competitors│ E              │
         X Y Z    │   Technical    │
                  │   Evaluation   │
                  └──────────────────┘
                  ┌──────────────────┐
                  │ Target    Value  │
                  └──────────────────┘
```

Figure 5.7 Details of phase 1: Product planning or product definition.

Review of QFD basics via the pencil example

We mentioned earlier that a well-developed house of quality is an essential ingredient for successful product development. Let us develop a house of quality for a pencil. The pencil example does not cover all details of QFD, but the example will give you a good understanding of QFD basics. The ideas developed in this example were germinated at a class on QFD at Hewlett-Packard, and the ideas have been validated by experts from the pencil industry.[3]

Listing of customer needs. The students acting as customers listed their needs. These needs or demands, known as the customer's voice, are tabulated in Table 5.1. Developing an excellent customer's voice is of the utmost importance, and we will share ideas on this subject later in this chapter.

Next, the house of quality, described in Fig. 5.7, was developed for the pencil. The contents of the figure were slightly rearranged; all the relevant factors, however, are included. The fully developed house of quality for the pencil is displayed in Fig. 5.8. The steps to develop the figure are described here.

TABLE 5.1 Customer's Voice or Demands for a Pencil

Writes	Nonsmearing eraser
Sharpens	Smells good
Does not roll	Lead does not break too easily
Easy to hold	Tells me type of lead in pencil
Does not smear	Writing is visible
Can erase	Can photo copy
Has choice in lead	Does not tear paper when I write
Has choice in color	No slivers
Is inexpensive	Not toxic
Eraser does not break	Soft for biting but does not break paint
Feels good	Sharpened point to last half page of text
Spells correctly	

Grouping of customer needs. The needs, gathered in Table 5.1, were grouped using the KJ method, which is named after Kawakito Jiro, who developed this tool. Refer to the discussion of the seven new quality control tools in App. 2. This tool uses the right, or creative, side of the brain to group verbal data. Moreover, the KJ tool retains the customer's own words in the house of quality as tertiary labels, such as "nonsmearing."

These groupings are given labels—for example, the secondary labels of "selection" or "quality," to succinctly describe the ideas captured in those particular groupings. Similarly, secondary labels are grouped to develop primary labels; for example, the secondary labels of "selection" and "quality" can be grouped under the primary label of "lead" (refer to Fig. 5.8). By deploying the tree diagram (one of the seven new quality control tools) on these primary and secondary labels, the unspoken needs of customers can be ferreted out.

Quality planning. The 10 columns in Fig. 5.8 under Quality Planning are developed. This incorporates competitive analysis, strategic positioning, and merchandising strategy. Let us focus on one demand, nonsmearing (sixth line under Customer's Voice). The claims column (column 1 under Quality Planning) contains zero for our example. Claims refers to the ability of our company to provide proper product warranty or service (claims).

The next column, rate of importance, is 5 on the scale of 1 to 5, 1 being the least important and 5 the most. This rating is provided by the customers during the customer interview phase.

The competitive evaluation columns (columns 3, 4, and 5) indicate that our company is rated 5 whereas both competitors X and Y are rated 2.

Since customers rated nonsmearing as an important need, the quality plan column (column 6) is targeted at 5, even though there is no threat from the competition. This means we plan for the best possible quality for this item of nonsmearing.

Quality Planning

Competitive Evaluation

Correlation Matrix

Legend
◎ = High Relationship
○ = Medium Relationship
△ = Weak Relationship

Customer's Voice

Primary label	Secondary label	Tertiary label	DURABILITY: Glue Strength	Eraser Bond Strength	Torque To Break Eraser	Torque To Break Point	Tolerance Wood Lead	Lead Off Center	PERFORMANCE: Darkness of Pencil Line	Lead Dust Generated	Lead Adhesion To Paper	Eraser Adhesion To Paper	FEELS GOOD: Weight	Balance	Hexagonality	Outside Diameter	Smoothness	SAFETY: Toxicity	Totals	One: Claims	Two: Rate of Importance	Three: Own Company	Four: Brand X	Five: Brand Y	Six: Quality Plan	Seven: Rate of Level Up	Eight: Sales Point	Nine: Absolute Weight	Ten: Quality Weight
LEAD	Selection	thickness				△		△		△	△		△			△				0	3	4	4	4	4	1.00		3.0	3.2
		hardness				◎			◎	◎	◎									0	4	2	2	3	4	2.00		8.0	8.6
		contrast							◎		○									0	3	3	3	4	4	1.33		4.0	4.3
	Quality	marks easily							△	○	◎									0	4	4	5	3	5	1.25		5.0	5.4
		marks legibly							○		◎									0	5	4	5	3	5	1.25	◎	9.4	10.0
		non smearing				**△**			**○**	**◎**	**◎**									**0**	**5**	**5**	**2**	**2**	**5**	**1.00**	**○**	**6.0**	**6.4**
		erasable				△			○	◎	◎									0	4	3	4	4	4	1.33		5.3	5.7
		point lasts				○			△	○										0	3	5	1	2	5	1.00		3.0	3.2
		doesn't break easily				◎	△		△							◎				0	5	5	2	2	5	1.00	○	6.0	6.4
ERASER		non smearing								○	△	◎						△		0	5	4	3	3	4	1.00	○	6.0	6.4
		does not break		△	◎							△								0	4	3	4	3	4	1.33	◎	8.0	8.6
PENCIL	Body	legible advertising													△		◎			0	1	3	5	5	5	1.67		1.7	1.8
		lead type													△		○			0	4	5	5	3	5	1.00		4.0	4.3
	Convenience	light weight											◎	○		○				0	3	4	4	4	4	1.00		3.0	3.2
		easy to hold											◎	◎	○	◎	◎			0	4	3	3	4	4	1.33	◎	8.0	8.6
		does not roll											△		◎					0	2	4	4	3	4	1.00		2.0	2.1
		sharpens to good pt	△				○	◎								○				0	5	4	4	4	4	1.00		5.0	5.4
		non toxic										△					△	◎		0	3	4	5	3	4	1.00		3.0	3.2
		inexpensive						○	△		△		○			○	△			0	3	3	3	4	3	1.00		3.0	3.2
		SUM	5.4	8.6	25.7	66.8	17.1	25.7	101.1	95.3	129.7	31.1	47.1	32.1	29.6	71.8	46.1	16.1	749.2									93.4	
		relative weight	0.7	1.1	3.4	8.9	2.3	3.4	13.5	12.7	17.3	4.1	6.3	4.3	4.0	9.6	6.1	2.1											

Figure 5.8 Product planning (pencil example).

Rate of Level Up (column 7) measures whether we are improving, maintaining, or decreasing quality—as compared with the competition. The Rate of Level Up is defined as: Quality Plan value/Competitive Rating value = 5/5 = 1.00 (that is, column 6/column 3). In this case, we are rated high and plan to stay high.

The planned sales points are illustrated in column 8. These are decided by management after studying customer needs and the competition. They will represent competitive strengths that can be conveyed in a sales and marketing campaign. The primary (or important) sales points are indicated by a double circle and carry a weight of 1.5, while the secondary sales points are indicated by single circles and each carry a numerical value of 1.2. These weights have been arbitrarily set, based on experience.

The absolute weight (column 9) is calculated by multiplying values of rate of importance, rate of level up, and sales point. Thus absolute weight = rate of importance × rate of level up × sales point. For non-smearing, the absolute weight = $5 \times 1 \times 1.2 = 6.0$.

All the absolute weights are added vertically (in column 9); here the sum is 93.4. Each of the absolute weights is normalized to a base of 100 to obtain quality weight, tabulated in column 10. The quality weight is calculated to be 6.4 for nonsmearing.

Purpose of numerical evaluation. Numerical evaluation helps give the team quantitative information with which to make correct decisions. Glancing at column 10, we can easily determine which are the important needs. We can show the importance in another way: Figure 5.9 shows the competitive data pictorially. The data here is collected from columns 3, 4, and 5 of Fig. 5.8. The picture makes it easy to understand our own strengths with respect to the competition; it is very easy to identify challenges and make a conscious decision to either address them or leave them alone.

Defining product characteristics. One method for developing product characteristics or design requirements of a product is *analytical brainstorming*. These characteristics are then organized, using KJ and a tree diagram, as was done with customer needs. The results are shown at the top of Fig. 5.8 with tertiary (for example, lead dust generated) and secondary labels (for example, performance). Primary labels were not developed for this example.

Analytical brainstorming is a term coined by the author to distinguish this specialized brainstorming from the generally used free-form brainstorming, where knowledge of subject matter is not a prerequisite. Engineering or design knowledge is needed to generate ideas to meet customer needs. The brainstorming is carried out se-

Daily Process Management 145

PICTORIAL COMPETITIVE EVALUATION -- FOR A PENCIL

		Customer's Voice	Customer's Eyes 1 2 3 4 5	
Lead	Selection	Thickness		
		Hardness		
		Contrast		
	Quality	Marks easily		
		Marks legibly		
		Non-smearing		
		Erasable		
		Point lasts		
		Does not break easily		
Eraser	Body	Non-smearing		
		Does not break		
		Legible advertising		
		Lead type		
Pencil	Convenience	Light weight		
		Easy to hold		
		Does not roll		
		Sharpens to a good point		
		Non-toxic		
		Inexpensive		

Key
O   Our own company
X   Company x
△   Company y

Note: This data is obtained from phase I. All circles, indicating our company's performance, are connected.

Strengths and weaknesses are immediately apparent, and decisions on how to respond in the marketplace can be made.

Figure 5.9 Pictorial competitive evaluation—developed from phase 1.

quentially with respect to the needs of customers, which is another attribute of analytical brainstorming.

Another way to generate design requirements is to use a cause and effect diagram. Note that the correlation matrix (the triangular roof area) between design requirements was not carried out for the pencil example. In real life this is an invaluable step and must be incorporated.

Developing relationships. Next, the relationship between customer's voice and design requirements is evaluated. This relationship is indicated by symbols, such as a triangle, circle, and double circle.

Once again we will develop this section by focusing on the nonsmearing quality of lead. When the QFD team sees no relationship, the cell is left blank. The team felt that the torque to break point and nonsmearing had a weak relationship, and so a triangle was placed in that cell. Darkness of pencil line and nonsmearing had a medium relationship according to the team members, so the cell contains a circle. The strong relationships between design requirements like lead dust generated, lead adhesion to paper, and nonsmearing received a double circle in the cell.

For our example weights of 1, 2, and 3 are assigned to a triangle, circle, and double circle, respectively. The numerical values in the relationship matrix are then derived by multiplying the quality weight and the numerical value of the symbols in each cell; these are added vertically, and the sum is entered in the sum cell, at the bottom of Fig. 5.8.

146 Chapter Five

All the columns are added this way. The sum for adhesion to paper is 129.7, and for torque to break point it is 66.8. These calculations help set priorities for design requirements that will generate increased customer satisfaction.

Developing the next phase. In the case of the pencil example, lead adhesion to paper received maximum weight. Therefore lead adhesion to paper is carried to the part deployment phase (phase II) as shown in Fig. 5.10. Other important characteristics, such as darkness of pencil line, could also be deployed to phase II. In phase II we apply a similar rigorous analysis as in phase I and set priorities for parts that are critical for increasing customer satisfaction. Graphite, clay, and gum are identified as these parts. The parts now guide the process planning phase—phase III.

The critical processes, for example, grinder pressure, are identified in phase III, to guide production planning in phase IV. A clear understanding of production knob settings (see Fig. 5.10) for the critical processes are thus established. Appropriate control charts are designed before manufacturing begins—specifically, a quality checkpoints document is prepared. Refer to the section "Quality Checkpoints Chart and System" later in this chapter for more details.

We have given you a glimpse of the steps taken to address the customer need of nonsmearing throughout the product life cycle in order to deliver that need to the customer. Finally, when the product reaches the market,

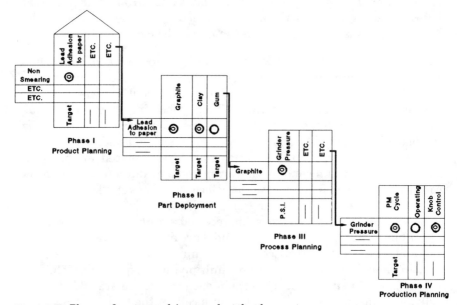

Figure 5.10 Phases of customer-driven product development.

the customer's voice should be updated and validated. This would complete the entire PDCA cycle for developing the next generation of pencil.

How many phases or matrices should you have? In Fig. 5.10, we have shown four phases. Other proponents of QFD suggest up to 30 (e.g., Yoji Akao). We suggest you begin with one initially: phase I shown in Fig. 5.10—also known as the house of quality. The house of quality is illustrated in detail in Figs. 5.7 and 5.8. Once you have mastered this phase, proceed to the others.

With this fundamental understanding of QFD, we would like to study two critical success factors in depth:

1. Developing the customer's voice
2. A process for increasing QFD success in your company

Developing the customer's voice for QFD

Who is the customer? It is critical to properly select the customer. This can be done via market analysis, market segmentation, and market potential.

Developing the customer's voice. After the customer is identified, the next step is to develop the customer's voice. Developing the voice of the customer is the most important and most difficult task. Since the voice of the customer drives all other activities of the development process, its importance is easy to comprehend. The voice may be spoken or unspoken, may be direct or indirect. One of the challenges is to combine many voices to glean the essence. It is critical to tap all sources of the customer's voice.

Some of the direct sources are input (words) from a customer (this could be accomplished using surveys or focus groups), repair records, replacement records of worn out or defective parts, identification of defects, enhancements or requests for enhancements, market research data, list of things done right in previous products, list of things done wrong in previous products, manufacturing problems caused by poor design, requests for engineering changes, supplier's difficulties in the past, and even a test engineer's voice, which could include the parameters that were not tested in previous designs. More indirect sources are observing customers' difficulties when using your product, observing the difficulties customers experience when they are solving their own work problems, capturing murmurs at trade shows, getting ideas from technical media, listening to some customers experiencing the product or service from their point of view, anticipating future needs of customers by capturing comments such as "I only wish...."

We would like to make one important point. Every one of us holds a set of paradigms by which we view the world around us. Our ability to hear what is being communicated is heavily affected by the paradigms we hold. If a customer's need does not fit our paradigm, we may fail to hear the need. In order to minimize missed customer needs, use as many tools as possible, including tape and video recorders. We recommend that you view a video tape by futurist Joel Barker called "Discovering the Future Business of Paradigms"[4] before contacting your customers. This will make you sensitive to the fact that we all have our own paradigms. This awareness will facilitate a conscious effort to capture the customer's voice.

Steps for developing a high-quality customer's voice
1. Collect information from every conceivable customer source. Many sources are listed in the previous section.
2. Express the voice in terms of need. Do not include what is not required by the customer.
3. Use the exact words of the customer whenever possible. The unfiltered expression of the customer's voice is critical.
4. Express customer needs in brief positive phrases with at least a subject and a verb and no double negatives.
5. List only one need per line.
6. Focus on the target market only.
7. Stratify types of customers when applicable. Software products, for example, have customers who are programmers, systems integrators, systems maintainers, managers who are decision makers, and end users. Needs will vary with the type of customer strata; the degree of importance will vary even if the needs are identical.

After all the needs are collected, a tool called an *affinity diagram,* or *KJ,* is very useful in grouping customer needs. This is one of the seven new quality control tools. The use of tools can help increase management participation in TQC activities.

A company invests in market research to be able to market products and services that would thrill current and future customers and result in increased market share. The objective of research is to develop deeper and more imaginative understanding of customer needs, but as explained in Noriaki Kano's *two-dimensional quality* model (Chap. 1), this is not a trivial effort. The KJ method allows a team to be more imaginative in capturing customer needs.

The KJ method should be used to organize the customer's voice. Being imaginative also facilitates capturing customers' emotions. Since the buying decision of customers includes emotions, it makes sense to include emotions in product development. In the traditional management environment predetermined categories are used to organize the customer's voice. But categories do not help capture emotions. Moreover, we are filtering the customer's voice through our preexisting bias, for example, by adding categories which are conceived by the designer rather than the customer, hence biasing the product definition. The unfiltered customer's voice has a higher probability of being included when KJ is used correctly.

Another very useful tool in the seven new quality control tools is the tree diagram. Unlike KJ, the tree diagram is a very logical and systematic tool. Using the tree diagram along with KJ helps the QFD team decipher the needs which customers may have neglected to mention. The topics developed when KJ is combined with the thoroughness of a tree diagram enable us to get at the unspoken words of the customer. Understanding and capitalizing on unexpressed needs can be a key competitive advantage. The QFD process steps will help you move toward this goal.

Recommendations for increasing QFD success in your company

We will give some guidelines for ensuring success in implementing QFD in your organization. During this discussion, refer to Fig. 5.11.

Choose a consultant or facilitator

Attributes. In Fig. 5.11, we show a flow chart for ensuring the success of QFD in your company. The consultant can be from an internal or external source. A good consultant is critical to the success of this effort. The consultant should have the respect of all team members and should be perceived as an expert by the team. The consultant should be analytical, understand QFD, believe in the quality-oriented management philosophy, and work well with people. The consultant must establish and meet timelines, work on a win-win philosophy, proceed only when team consensus is established, know how to facilitate successful meetings, and build well-functioning teams.

Education. There are some fundamental educational requirements for the consultant. TQC philosophy and PDCA cycle should be an integral aspect of his or her education. The consultant must recognize times when experts from other disciplines are needed by the QFD team. A thorough understanding of QFD is assumed.

Role of the consultant or facilitator. A consultant must be a fast thinker and two steps ahead of the team. The person must be a process expert

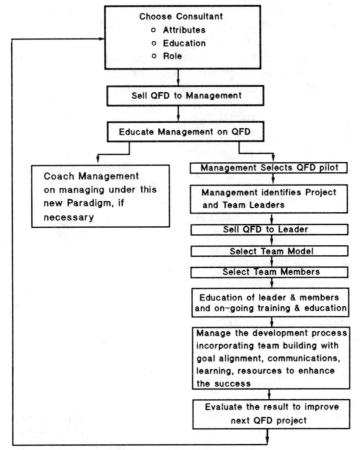

Figure 5.11 Process flow for increasing QFD success.

and help the team stay focused on the task. A refresher, just in time training session for the QFD team given by the consultant will alleviate the stress of not remembering details learned in the past. Training should directly address the task at hand.

It is well known that venture capitalists do not invest in an idea; they invest in a person who has the idea. The team's confidence in a consultant will increase prospects for success. The consultant also functions as a change agent when QFD is initiated in the traditional business environment. After the transformation has taken place and TQC philosophy is pervasive in an organization, the significance of the consultant will diminish.

Sell QFD to management. A package to sell QFD to management is presented in the paper "Does Success Have a Secret?"[5] You may want to consult the paper for more ideas on selling. Managers tune into economic benefits, so these benefits should be explained with data. Alternatively, you can use the data on why products succeed and fail from Chap. 2 and the benefits of QFD listed earlier in this section.

Educate management on QFD. This session should contain the fundamentals of QFD such as the pencil example described earlier. In addition, the session should answer any questions on timeline, resources, and the learning curve. All the benefits cited earlier in the chapter may not be achieved for the first pilot project.

This session should also develop a realistic expectation of resources needed for success. This education is a combination of selling the concept, developing realistic expectations, and obtaining a firm commitment by management. An honest communication that return on investment (ROI) will not be instantaneous should be given. Part of the problem is that many businesses and companies do not know how to connect the cost of losing customers or market share with generally acceptable internal accounting figures. The QFD process, however, will give management a clearer understanding of what went wrong. More importantly, it will help identify how to improve the next time by eliminating root causes of process malfunction. Today an individual designer or manager is blamed for failure, but no steps are generally taken to eliminate recurrence of the failure in the future.

Coach management. Correct management behavior will enhance the chance of success tremendously. Some managers will be able to adopt this behavior easily, but others will need coaching. Here are some examples of desirable behavior: Listen to the team's progress by attending reviews and offer support by facilitating or reallocating resources needed for the success of the QFD team. Do not discourage team members by asking "Why are you doing such an such?" or saying "Do this and that." Instead the comments should be "What did you learn from doing this?" or "How will you use this information for your project?" Management should encourage documenting things done right as well as those done wrong; this will create a more open environment which is free of blame and focuses on learning from the process.

Management selects QFD project. A high-level management team should be involved with the project selection process for QFD. Management should reach consensus on the selection. The project should be a product (or service) critical for the success of the business. Correct project selection will increase the probability that good market research is available or will be funded for the project. Good market research is a critical starting point for a QFD project. Management should also identify the team leader.

Sell QFD to the leader. The same process of selling the concept and developing interest should continue to the level of project leader and team members. A successful QFD project will require hard work by the leader and team. Management should not demand that QFD be used; the team should be sold on the idea.

Team models. There are many open questions in implementing QFD. One question is what is the best model for team structures? Here are some possibilities:

- An assigned leadership with a core team that has members from all functions for the duration of the product development cycle.
- Dynamic membership and fluid leadership that are in tune with the product developmental phase of the cycle. In other words, in the beginning when customer's voice is being developed, the team is led by marketing and the members are from the next phases, R&D and manufacturing. The leadership would then switch to R&D with membership from the previous and next links of the process, that is, marketing and functions involved with manufacturing.

An organization should study the question of team structure with appropriate data to adopt the best model for their culture and product type (for example, a simple product or a complex product).

Criteria for selecting members. Here are some guidelines on selecting members for your first QFD project:

1. Members should have approximately equal rank to minimize position power influence and to facilitate good communication, exchange of ideas, and team dynamics.
2. Members should have a sense of adventure.
3. Members should be calculated risk takers.
4. Members should be open to new ideas.
5. Members should be learning and growing.
6. They should not be too vested in the established procedures or thinking that is how things are done.

Six to eight people constitute a team that functions well in terms of size. Should every expert be on the team or just those charged with the responsibility of developing the product? The second alternative has merits provided we bring in experts to address the special needs of the team as they arise. Having limited representation from each critical functional area will ensure that the representatives assume responsibility to obtain pertinent information.

Who should be on the QFD team? The membership comes from the areas listed below. This is a long list and priorities will have to be made. During the life of the team, however, many of the personnel listed will have to be contacted.

1. Functions that are in touch with the customers
   - Marketing: product or technical
   - Sales: telesales, direct sales, catalogers
   - Service: repairs, personnel connected with warranty management
   - Hot (problem) site personnel
   - Personnel connected with visits to customers' sites and various company sites
   - Market researchers
2. Functions involved with design
   - Design engineers
   - Technical researchers
   - Academic experts in a specific technical field
   - Test engineers
   - Process designing engineers
   - Cost engineers
   - Experts who design experiments for scientific study of breakthrough engineering or tradeoff effects
3. Manufacturing functions
   - Manufacturing line personnel
   - Maintenance personnel of manufacturing equipment
   - Manufacturing process engineers
   - Packaging engineers
   - Material engineers
   - People in charge of keeping the manufacturing process under statistical control

Training and education. Training the team is very important. Everybody should learn the correct vocabulary; communications improve as a result of common language, which in turn helps with team dynamics. Goal alignment is achieved since all team members focus on customer needs as they carry out their own tasks and activities in order to develop a dynamite product which will thrill the customer. This goal alignment also helps with team building. The 2-day training provided by organizers such as ASI will be adequate.

What else is needed for success?
1. A realistic timeline
2. Regular project reviews
3. Milestone celebrations to keep interest high and to develop a sense of closure
4. Recognition from management
5. Sharing with other teams to facilitate deeper learning

Summary of QFD

We have discussed the benefits of QFD. A pencil example has been used to describe the key steps for developing QFD tables. Ideas for developing a high-quality customer's voice and a flow chart to increase QFD success has been shared. Three success factors are an accurate customer's voice, strong management commitment, and a good consultant. In closing we would like to quote Chris Fosse of Omark Industries: "QFD does not give you the answers. It is a tool to organize thinking and activity to help you ask the right questions."

Design Standards

We discussed the importance of and need for standards in Chap. 4. These standards embody the best practices which have proven to be effective. Repeated use will ensure better and predictable results. The standards we will list here should be used in every product during the design stage. With time, based on available software, many of the standards can be incorporated into design tools such as computer-aided design software. Design engineers must be trained in the use of these standards, and, in addition, their application should be mandatory. The standards listed here are the ones we estimate will have the greatest impact on product reliability for electronic and electrical products.

For other types of products, you will need to select and define applicable standards. The failure mode effects analysis (**FMEA**) process, however, will be applicable for all products (discussed later in the chapter).

Thermal design and measurement

Most failure mechanisms in electronic and electrical products are caused or accelerated by heat. In fact many of the failure mechanisms in electrical and electronic components *double* their failure rate with an increase of only 10°Celsius. Hence, paying proper attention during the design stage to power dissipation and heat transfer by conduction or convection within a product can minimize such failures.

What needs to be done is first to calculate and analyze the power dissipation and heat transfer of components using design rules and then to confine the temperature within recommended manufacturer's limits. Next, thermal measurements are required to check for and verify proper thermal operation. This can be done with both infrared imaging equipment and contact transducers. Ensuring both proper thermal design and measurements can eliminate thermal hot spots and heat problems that induce failures. If you do not have a program to ensure thermal design and measurements, the recommended reading in the Bibliography will provide further direction.

Component derating

Another major cause for failure in electrical and electronic products is the failure of individual components because of high stress. Such failures can be reduced or eliminated by component derating. We define derating as the limiting of the level of stress applied to the component so that the maximum stress in service is well within the component's specified or demonstrated maximum capabilities in order to enhance its reliability. For example, a reduced power level in a semiconductor device results in a lower junction temperature. Similarly, reduced voltage across a capacitor results in a lower electric field in the dielectric, thus enhancing reliability.

Where a particular stress level is required, derating can also be accomplished by specifying or selecting a component which is more robust to that stress. For example, at the same power level, the use of a semiconductor package designed for a lower thermal resistance will result in a lower junction temperature. Similarly, use of an electrolytic capacitor with a larger surface area will result in a lower electrolytic temperature, if other characteristics (such as ripple current and equivalent series resistance) are equal.

Another derating method is a reduction in the expected component performance. For example, a reduction in applied clock frequency may allow the use of a wider temperature or voltage range. Similarly, use of a bipolar transistor at 50 percent of the minimum specified gain will allow operation at lower temperatures.

Proper derating methods will result in lower failure rates because stress levels on components are reduced. Component derating is one of the most powerful methods available to enhance the reliability of the equipment in which the components are used.

After the design, application, and derating, the designer should follow a stress analysis program which includes voltage, current, power, and junction temperature. The analysis should include the influence of the geometry (layout) of the components on the printed circuit assembly and the arrangement of the assembly within the system. The thermal measurement methods discussed in the previous section are also recommended. Hence, component derating and thermal design and measurement are methodologies that complement each other.

For the customer, enhanced reliability from component derating will result in increased uptime, reduced distractions because of service calls, decreased costs, and increased satisfaction. More details on component derating are available in the suggested readings listed in the Bibliography.

Specific electrical and mechanical standards

We recommend that you set up specific electrical and mechanical standards in areas where they don't exist. For example, if your products are portable, usually in plastic cases, you should set up documented standards based on your experience for such attributes as the type of plastic material to be used for different applications (outdoor use, indoor use, luxury finish, and so on), recommended thickness of cases, methods of attaching metal or other material to the cases, and methods of sealing (for waterproof applications) and closing the cases.

Possible areas for generic standards include printed circuit boards, displays, electronic and electrical circuits (which can be used in different products), mechanical devices and structures, and so on. Many of these are available from industry associations such as EIA, IPC, ASTM, and ASME. Safety and regulatory standards are available from organizations such as VDE, UL, CSA, and FCC.

With time, as you build up these documented standards, it will become easier to train new designers to build high-quality products. Good internal standards that document a wealth of experience can become a competitive advantage in current product offerings and in new applications of current technologies.

Supplier management

Approved or certified parts. Developing a good working relationship or partnership with suppliers and certifying or managing them is essential for success. This includes ensuring a supply of approved or certified parts.

The use of approved or certified parts by designers can help reduce design time. An approved or certified part is one which consistently meets certain standards. If the designer has access to a database of such parts, he or she can select a part for a new design with the assurance that the part will consistently meet certain specifications and be reliable within those specifications.

In addition, the certified supplier will be a good business partner to your company. More important, there will be reasonable assurance that the approved part will cause no problems in the new design. The alternative would be to seek a new part and a new supplier—in such a case additional time and testing is needed to ensure reliability. Hence a good supplier management program can indeed reduce design time. One of the keys to success here is to have a good supplier management and certification program. Certified parts can move straight from the supplier to the manufacturing line. There are numerous texts available on supplier management. We will not discuss this topic but urge you to review the literature on the subject, if you don't have a good program.

Failure mode effects analysis

In Chap. 4 we discussed the problem-solving hierarchy. Ultimately you should be in the third stage—problem prediction and prevention. One of the tools that can help do this is FMEA, which is a very effective method of design reliability analysis. This method originated with National Aeronautic and Space Agency (NASA) engineers and is specified in United States Military Standard MIL-STD-16291 (Procedures for Performing a Failure Mode, Effects and Criticality Analysis).

This is not a commonly used method, so we will give you an idea of what it is, but further details are best obtained from the readings listed in the Bibliography. The following discussion on FMEA comes from Rajan Vrittamani of Hewlett-Packard, to whom we are deeply indebted.

Definition and benefits of FMEA. FMEA is a systematic process to evaluate failure modes and causes associated with the design and manufacturing processes of a new product. It is somewhat similar to the potential problem analysis phase of the Kepner-Tregoe program.

The FMEA process. A list of potential failure modes of each component or subassembly is made, and then each mode is given a numeric rating for frequency of occurrence, criticality, and probability of detection. Finally, these three numbers are multiplied together to obtain the risk priority number (RPN), which is used to guide the design effort to the most critical problems first. Typically only the high RPN numbered items (say 50 percent of total items) are eliminated. Refer to Fig. 5.12, where we show a simple flow chart of FMEA activity.

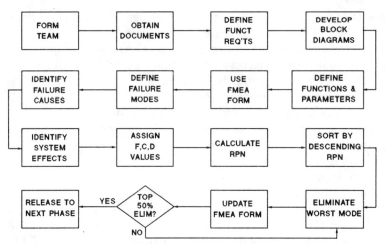

Figure 5.12 Flow chart of FMEA activity.

Two aspects of FMEA are particularly important: a team approach and timeliness. The team approach is vital because the broader the expertise that is brought to bear on making and assigning values to the failure mode list, the more effective the FMEA will be. Timeliness is important because FMEA is primarily a preventative tool which can help steer design decisions between alternatives. Hence FMEA should be started early in the design stage. FMEA is equally applicable to hardware or software and to components or systems.

The RPN calculated by FMEA allows us to set priorities for the failure mode list, guiding the design effort to the most critical areas first. It also provides a documentary record of the failure prevention efforts of the design team. This will be helpful to management in gauging the quality and extent of the effort, to production in solving problems which occur despite these efforts, and to future projects which can benefit from all the work and thinking which went into the failure mode and cause lists.

Benefits of FMEA. Eliminating potential failure modes has both short- and long-term benefits. The short-term benefit is most often recognized because it represents savings of the costs of repair, retest, and downtime. The long-term benefit is much more difficult to measure since it relates to the customer's satisfaction with the product and his or her perception of its quality. That perception affects future purchases of products and is crucial in creating a good image for your products.

FMEA supports the design process. FMEA also strengthens and supports the design process by:

- Aiding in the selection of alternatives during design.
- Increasing the probability that potential failure modes and their effects on system operation have been considered during design.
- Providing additional information to aid in the planning of thorough and efficient test programs.
- Developing a list of potential failure modes ranked according to their probable effect on the customer. These failure modes can then be minimized or eliminated by design effort.
- Providing an open, documented format for recommending and tracking risk-reducing actions.
- Identifying known and potential failure modes which might otherwise be overlooked.
- Detecting primary but often minor failures which may cause serious secondary failures.
- Detecting areas where "fail safe" or "fail soft" features are needed.
- Providing a fresh viewpoint in understanding a system's functions.

Requirements to perform FMEA

1. A team of people with a commitment to improve the ability of the design to meet customer needs.
2. Schematics and block diagrams of each level of the system, from subassemblies to the complete system.
3. Component specifications, parts lists, and design data.
4. Functional specifications of modules, subassemblies, etc.
5. Manufacturing requirements and details of the processes to be used.
6. FMEA forms (paper or electronic) and a list of any special considerations, such as safety or regulatory, that are applicable to this product. Refer to Fig. 5.13 for a sample FMEA form. The form has been completed for a portion of a medical product used for patient monitoring. If during design—after setting priorities—we decided to prevent only one failure mode, we would select the mode with the highest RPN number.

Summary of FMEA. We recommend the use of the FMEA technique for all product design, preventive maintenance schedules, and test equipment. Initially, to get started, you can use this in products that have potential liability problems, for example, power supplies, fuel equipment, products that cannot be easily tested, totally new designs, or test equipment. With time this should be a standard design for reliability and quality technique.

FAILURE MODES AND EFFECTS ANALYSIS

Product: __MAGIC__ Assembly: __XMTR RF Module__
Engineer: __Rajan__ Date: __2/29/89__

Assy ID	Functional Block	Failure Mode	Failure Cause	Effect of Failure on System	F	C	D	RPN	Corrective Action or Remark
1	Oscillator tank circuit	No output frequency	Drift trim cap Defrect Induc, cap, or Q1 Shorted res	No RF sig from XMTR error msg on the screen	2	4	1	8	Error message to be defined. Trim caps should be sealed with epoxy
		Wrong Freq. Wrong Dev. level	Drifted trim cap						
2	RF tank circuit	Output pwr level low	Def transistor Drifted trim cap or pot	Weak patient sig possible bit errors in that channel	2	3	3	18	Invalid data message on the screen
		Output pwr level high	Shorted CR3 Shorted res						
			Drifted trimpot Open CR3	High current drain from XMTR battery	2	3	5	30	Trim pots should be sealed with epoxy

KEY:
F: Frequency
C: Criticality
D: Detection
1 = Remote 2=Low 3=Moderate 4=High 5=Very High
RPN: Risk Priority Number = FxCxD

Figure 5.13 A completed FMEA form.

Analysis of the New Product Development Process via the Project Postmortem and T-Type Matrix

Problems in manufacturing can be detected, analyzed, and resolved relatively easily for several reasons: Good process documentation and deviation analysis are usually available (refer to the section "Quality Assurance"), problems are stereotyped, and problems occur during a short manufacturing stage.

In new product development, however, the relationship between problem generation and problem detection is obscure. In addition, the new product development stage can take a long time, ranging from months to several years. Because of these issues, problems are often resolved on an immediate ad hoc basis with little concern for the root cause of the problem, which could have occurred years ago.

Typically, about 50 to 80 percent of problems discovered in the later stages of manufacturing and at the customer's location are created in the early stages of product development. This means that if we catch problems early, or better still eliminate problem generation, we will save the time, effort, and resources needed to fix the problem later. Customer satisfaction will also increase.

We provide here a discussion on how the new product development process can be analyzed, with the purpose of reducing design changes and problems and hence providing continuous improvement. We offer two specific methodologies that will help:

- The project postmortem
- The T-type matrix for more detailed analysis

The project postmortem

After a product has been designed, manufactured, and delivered to the customer, a review should be conducted to analyze the product in terms of market acceptance, the design process, costs, ease of manufacturing, and quality. Such an analysis, or postmortem, is useful to understand the product development process. The knowledge gained will be used to improve the entire product development process from collecting customer needs, designing and manufacturing the product, delivery, and usage by the customer. This postmortem is done after product release (see Fig. 5.5a). Let us go into some detail and review an actual example.

Postmortem objective and benefits. *Objective:* To analyze the product development cycle after product release in order to understand strengths and weaknesses. The knowledge gained will be used to improve future product development cycles. The postmortem process will:

- Strengthen cooperation of product development teams
- Facilitate sharing of experience and learning throughout an entity
- Allow understanding of the root causes of problem generation and design changes in new product development
- Reduce future product development time—measured by time to market, break-even time, and so on
- Reduce future start-up costs and delays caused by redesigns and

production changes after product release to manufacturing or the customer
- Improve product quality and customer satisfaction for future products
- Ensure a more efficient product development process

Postmortem process. Typically, a postmortem or project review team is selected about 2 months after the initial product release. The team includes representatives from the R&D, marketing, manufacturing, and quality departments. It could be led by the project leader (of the product under review), a production manager, or a quality assurance manager. It will meet several times and discuss the issues shown in the report below. All of the required data will be available during the 3-month period after product release. Much of the R&D process data, however, will be available at product release; hence, this data can be collected earlier by R&D.

Postmortem Process Flow

	Process step	When done
1.0	Project released to manufacturing and shipped to customers	Product release = T
2.0	R&D reviews design process Use of T-type matrix	T + 1 month
3.0	Production engineering or quality assurance appoints postmortem task force leader	T + 2 months
4.0	Postmortem meeting planned Agenda Premeeting questionnaire, (request for data mailed out as per recommended format)	T + 2 months
5.0	Meetings take place Data collected and discussed for each item in recommended report format Deviations analyzed Problems reviewed	T + 3 months
6.0	Issue report Standard format Analysis and proposals	T + 4 months
7.0	Review proposals for corrective action to improve processes and practices Final meeting with senior management Agreement on proposals Short- and long-term plans	T + 5 months

Note: Step 2 is driven by the R&D department, while steps 3 to 7 are driven by a task force leader.

Project postmortem—report format. The following is an outline of a recommended format. Note that the specific items measured at different companies will vary with product type, especially for items 7 and 8 below.

1. *Project Identification.*
   - Product name and number.
   - Team members.
2. *Objective.* Standard "canned" phrase given in the postmortem objective shown earlier.
3. *Marketing.*
   - Forecasts versus actual, and deviation analysis.
   - Orders and shipments.
   - Customer acceptance, customer feedback and comments.
   - Planned profit versus actual, and deviation analysis.
   - Proposal and corrective action.
4. *Research and development (R&D).*
   - Design time, planned versus actual, and deviation analysis.
   - R&D process analysis, that is, design changes and rework required (redesign). A design process analysis using a T-type matrix is recommended. The T-type analysis matrix is discussed next in this section.
   - Proposal and corrective action.
5. *Manufacturing.*
   - Costs, planned versus actual, and deviation analysis.
   - Manufacturing difficulties caused by design.
   - Number of design errors caught in manufacturing, that is, production changes that were made to correct for design errors. The T-type matrix in item 4, above, will also provide some of this data.
   - Supplier problems or issues.
   - Material and parts issues.
   - Production yields and scrap—planned versus actual and deviation analysis. Proposal and corrective action.

6. *Quality assurance.*
   - Test data.
   - Failure data.
   - Customer feedback, comments, and issues.
   - Proposal and corrective action.
7. *Design error performance measures.* This is to compare design errors between this project and past projects. The comparison gives an indication of improvement over time.
   - Rate of design changes (measures production changes after product release versus past trends).
   - Rate of first run defects (measures errors during the first prototype run versus past trends).
   - Proposals and corrective action.
8. *Standardization issues.* The purpose here is to review progress of automation and purchase of approved parts versus past trends.
   - Rate of automation (measures percentage of automated assembly versus past trends).
   - Rate of use of approved or certified parts versus past trends.
   - Proposal and corrective action.
9. *Overall summary.*
   - Summary of learning points.
   - What will be changed and corrective action to improve current processes.
10. *Appendix.*
    - Table of contents.
    - Supporting data for all items discussed in main report, that is, items 3 to 9.
    - Include long-term trend data for items 7 (design errors) and 8 (standardization goals).

Postmortem meeting guidelines. How should a postmortem meeting be conducted? Here are some guidelines, based on our experience, on how to ensure productive and effective postmortem meetings.

1. Prior to the first meeting, the team leader must send out a detailed questionnaire. The questionnaire should be very specific and should

be based on the report format proposed earlier. The participants should be given enough time to collect the needed data.

2. At postmortem meeting remember to:
   - Identify successes and give credit.
   - Listen to everybody.
   - Prioritize problems.
   - Identify weaknesses, analyze for root cause, and then propose corrective action.
   - Keep good minutes.
   - Stick to the facts, use data.

   And do not:
   - Blame anybody; everybody is there to learn.
   - Divide up into competing groups.
   - Turn the meeting into a complaints session.
   - Deviate from the agenda, allow story telling, or "dog and pony" shows.
   - Assign action items to those who do not accept them.
   - Try to solve all problems.

3. After the project postmortem, remember to:
   - Distribute the report.
   - Share conclusions and get acceptance for proposals from senior management.
   - Keep performance measures on postmortem process at your entity, such as:
     - Percentage of projects with postmortems.
     - Timeliness of postmortem activity.
     - A list of and number of contributions to the best practice file or standard update reports.
     - A list of and number of problems that recur from one project to another.
   - Follow-up on proposed corrective actions, that is, were they implemented? This is important.
   - Ensure that the lessons learned are institutionalized.

166 Chapter Five

For the last three items, we suggest you have a central custodian for the process, for example, a R&D standards manager or a neutral quality assurance manager.

Sample project postmortem report. We show a portion of a report in Fig. 5.14a and 5.14b. Note the summary and recommendations. These will be used to improve the design process in the future.

PROJECT POSTMORTEM   Page 1 of 15

Product: 5000X Components Tester   2 August 1990

①Team Members:
    R&D Section Manager
    Marketing Manager
    Production Manager
    Engineering Section Manager

5000X Project Postmortem, Page 2 of 15

3. MARKETING REVIEW

  A) Forecast & Shipments

②Objective

    The 5000X Components Tester has been well received by Customers. Sales, however, are at 71% of forecast during the last 6 months. The breakdown is as follows:
    * Sales to Market Segment A (off-line testing) = 120% of forecast
    * Sales to Market Segment B (in line testing) = 43% of forecast
    * Overall Sales are at 71% of forecast (Refer to appendix for more data)

  b) Profits

    The operating profit year to date is 8.3%, which is below forecast and is due to shipments below forecast (Refer to appendix for more data)

  c) Sales Deviation Analysis:

    Although we have exceeded forecasted Sales in Market Segment A, Sales to Market Segment B are below forecast due to Customer dissatisfaction with the feature set. The 5000X is not very appropriate for certain in-line testing. We interviewed customers in both market segments.
    All the needs of market segment A & B were captured in the top level QFD table (5000-A1), but in the second revision (5000-A2), some needs were dropped because of possible complexity and project delay. The decision to drop was made by the R&D engineer in the absence of the section manager. This severely impacted the needs/requests of Segment B/ In line customers.

Action: Establish a formal procedure for discarding any customer request.

Figure 5.14a A portion of a project postmortem report.

```
                    5000X Project Postmortem, Page 9 of 15

8. Standardization Issues
   * Rate of Automation:
     This is defined as  number of fully automated component insertions
                        ─────────────────────────────────────────────── X 100%
                              number of total components
     The rate for the 5000X was 73%, which compares favorably with the
     division average of 65% and 1987 target of 70%.

   * Rate of use of Approved/Certified parts is defined as
         number of approved parts used
         ───────────────────────────── X 100%
           number of total parts used
```

```
                                5000X Project Postmortem, Page 10 of 15
   9. Overall Summary
      Recommenbdations

      1) During collection and collation of Customer needs & requests, (in
         the QFD top level table) none shall be discarded without prior
         authorization by the Marketing and R&D Managers.

         Action:  a) Inform all Marketing and R&D staff
                  b) Add checklist in preliminary stage and Final
                     design review to discuss all customer
                     requests proposed for deletion
                     a) by Marketing and R&D section managers
                     b) by Marketing manager

      2) We conducted a FMEA ergonomic study for the 5000X tester front
         panel. The resulting instrument panel received rave reviews by
         customers. In the future we will conduct a FMEA study on all new
         product front panels.

         Action:  Quality Manager

      3) Paco was selected specifically as a supplier for the 5000X product.
         However, Paco was selected by R & D without
         consulting the Purchasing Department. Paco is a consumer electronics
         supplier and we have had numerous problems with them not being able
         to meet our specifications. In future, we must adhere to the policy of
         selecting suppliers through the Purchasing Department.

         Action : Purchasing Manager and R & D Manager
```

Figure 5.14b A portion of a project postmortem report.

The T-type matrix

The T-type matrix is developed from one of the matrices proposed in the seven new management tools. Much of the discussion here originates from the paper by Noriaki Kano.[6] The example, however, comes from Hewlett-Packard Singapore.

Objective of using the T-type matrix. The T-type matrix can be used to identify the root cause of problems or design changes that are generated in new product development. Proper use of the matrix will help in the following areas:

- Understand why the problem was generated and resulted in a design change.
- Understand why the problem was not detected when it was generated.
- Propose where and how we can improve our design process so as to eliminate or diminish such problems or design changes in the future.

Mechanics and Preparation of the T-type matrix. After a new product is designed and released to manufacturing, data on design changes (anything requiring rework) is collected and displayed in a T-type matrix. For each design change or problem the data is displayed in phases, which are:

> Phase A: The problem-generation phase, that is, when the design error was created.
>
> Phase B: When the problem could have been discovered phase. This can be during phase A or later.
>
> Phase C: The actual problem detection phase. This can be in phase B or later.

Look again at the three phases listed. Obviously, we never want problems or errors to occur. And if they do, we want to detect them immediately, in the problem-generation phase. Alas, this cannot happen every time; hence we need to identify the three phases of a problems or design change. The three phases (A, B, and C) are displayed clockwise in the T-type matrix in Fig. 5.15.

Completing the matrix

1. For a new product that is being developed collect all problem and error data as and when they occur. Such errors can include a design change, a rewritten specification, problems detected in the prototype stage, all the way to problems detected after purchase by the customer. This data should be recorded in a central record book or computer file, which is accessible to the entire marketing, design, and manufacturing team. This activity calls for a cooperative effort by the entire team.

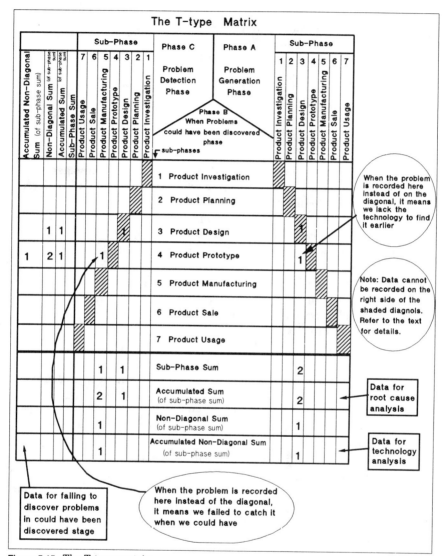

Figure 5.15 The T-type matrix.

2. We will use the following nomenclature:

   n = the maximum number of subphases in the new product development. In Fig. 5.15, we show seven.

Phase A_i = the actual subphase (i = 1, 2, 3, etc.) in Phase A (problem-generation phase).
In our example there are seven subphases in each of Phases A, B, and C

Phase B_j = the actual subphase (j = 1, 2, 3, etc.) in Phase B (could-have-been-discovered phase)

Phase C_k = the actual subphase (k = 1, 2, 3, etc.) in Phase C (problem-detection phase)

When we fill in the data we will display all the problems generated in phase A. We will also determine when the problems were detected—immediately or later.

3. First, complete the left-hand side of the matrix in Fig. 5.15. We call this area phase BC, and it denotes the relationship between phases B and C. Here we enter the following data: subphase BC, which means problems discovered in subphase C but which could have been discovered in subphase B. There are three possibilities: A problem discovered after generation, a problem discovered during generation, and a problem discovered before generation. Clearly, the last phase is impossible.

   a. Phase BC, where j < k. This means a problem was discovered in subphase C, later than when it could have been discovered in subphase B. For example, looking at Fig. 5.15, a problem was discovered in the manufacturing subphase 5 but could have been discovered in product prototype subphase 4. In such a case, put a 1 (for one problem or error) where subphase 5 (in phase C) intersects subphase 4 (in phase B) on the matrix. We have done so in Fig. 5.15.

   b. Phase BC, where j = k. This means a problem was discovered in subphase C at the same time that it could have been discovered in subphase B. For example, looking at Fig. 5.15, the problem was discovered in the design, subphase 3, and could have been detected in design, subphase 3. In such a case, put a 1 (for one problem or error) at the intersection, where subphase 3 (in phase C) intersects subphase 3 (in phase B). We have done so in Fig. 5.15. In this condition of j = k, you will be filling in the diagonal (the shaded boxes) in Fig. 5.15. This is the desired condition, where problems are detected when they can be discovered.

   c. Phase BC, where j > k. This means a problem was discovered in subphase C before it was made in subphase B. Clearly this cannot happen. An example is discovering a problem during product design that only the customer could discover after use. Hence, there will be no data entered on the right of the shaded diagonal in Fig. 5.15.

4. Next complete the right-hand side of the matrix in Fig. 5.15. We call this area phase AB, and it denotes the relationship between phases A and B. Here we enter the following data: subphase AB, which means problems that were generated in subphase A but were discovered in subphase B. There are three possibilities: a problem discovered after generation, a problem discovered during generation, and a problem discovered before generation. Clearly, the last phase is impossible.
   a. Phase AB, where i < j. This means the problem could have been discovered in subphase B after it was generated in subphase A. For example, looking at Fig. 5.15, a problem could have been discovered in the prototype, subphase 4, but was generated in the design, subphase 3. In such a case, put a 1 (for one problem or error) where subphase 3 (in phase A) intersects subphase 4 (in phase B). We have done so in Fig. 5.15.
   b. Phase AB, where i = j. This means a problem could have been discovered in subphase B at the same time it was generated in subphase A, for example, a problem generated in design, subphase 3, which could have been discovered in design, subphase 3. We have put in a 1 (for one problem or error) at the intersection of both design subphases in Fig. 5.15. In this condition i = j, you will be filling in the diagonal (the shaded boxes) in Fig. 5.15. This is the desired condition in the matrix, where problems are discovered as they are created.
   c. Phase AB, where i > k. This means a problem could have been discovered in subphase B before it was generated in subphase A. Clearly this cannot happen. Hence there will be no data entered to the right of the shaded diagonal.
5. Sum up the rows and columns in Fig. 5.15. There are four types of summing up; these are listed in the last four lines at the bottom and extreme left of the figure. The following are the four types of sums:
   a. Subphase sum: This is the sum of the problems in each subphase. Each phase is summed vertically (for phase A and C) or horizontally (for phase B).
   b. Accumulated sum (of subphase sum): This is the accumulated sum of the number of problems in the subphases, indicated in item (a) above. As we move from subphase 1 to subphase n in the subphase sum boxes, we add the numbers and put each addition in the accumulated sum box just below.
   c. Nondiagonal sum (of subphase sum): This is the sum of the problems in each subphase, excluding the problems in the diagonal boxes. These sums are the same as the number in (a) above but minus any numbers in diagonal shaded boxes.

d. *Nondiagonal accumulated sum (of subphase sum):* This is the accumulated sum of the nondiagonal sum [item (c) above]. As we move from subphase 1 to subphase n in the nondiagonal sum boxes, we add the numbers and put each addition in the box just below.
e. If necessary, we can also convert the summed data into percentages of all problems that occurred in that phase and show it in the same box by dividing each box into two (the percentage is in the lower portion of the box here). This is optional and is shown in Fig. 5.16. The data is converted into percentages to facilitate communication to management.

Analysis and interpretation of data. Analysis and interpretation will be segmented into three categories: why the problem was not discovered when it should have been, what test technology is needed to discover problems, and getting to the root cause of problem generation:

1. *Reason for not discovering a problem in the could-have-been-discovered phase.* Here we want to understand why the problem was not discovered when it could have been discovered, that is, why did we fail to catch it when it occurred? This data is collected from the left-hand side of the T-type matrix, specifically from the accumulated nondiagonal sum data. Refer to Fig. 5.15. The data can be displayed in the categories shown in Table 5.2. A Pareto diagram analysis should be prepared, followed by a plan to reduce the major causes of the problems. (We will give an example that will make this clear.)

2. *Lack of evaluation or test technology.* Here we want to understand why the problem could-have-been-discovered phase (phase B) occurs later than expected, later than the problem-generation phase (phase A). In other words, why wasn't the problem discovered when it was created? The reasons are typically lack of evaluation or test technology.

The data is collected from the right-hand bottom of the T-matrix—specifically from the accumulated nondiagonal sum data. Refer to Fig. 5.15. The data can be displayed in the categories shown in Table 5.3. A Pareto diagram analysis should be prepared, followed by a plan to reduce the major causes of the problems.

3. *Root cause of problem generation.* Here we want to understand why the problems are generated in the first place. The data from phase A, the problem-generation phase, is collected from the right-hand bottom of the T-type matrix, specifically the accumulated sum data. Refer to Fig. 5.15. It is then analyzed for root cause. The data can be displayed in the categories shown in Table 5.4. A Pareto diagram analysis should be prepared, followed by a plan to reduce the major causes of the problems.

TABLE 5.2 Common Causes for Not Discovering Problems

A. Checked but missed catching the problem
   A.1 Checkers mistake (careless, inexperienced)
   A.2 Inappropriate criteria
   A.3 Inappropriate checking methods
   A.4 Inappropriate item to check
   Note: Item A.1 is useful in separating human error from incorrect criteria or methods.

B. Did not check for problem
   B.1 Checking judged as unnecessary
   B.2 New problem, never planned to check
   B.3 Impossible to check for

TABLE 5.3 Common Causes for Lack of Test Technology

C. Technology exists inside the company or division.
   C.1 Insufficient time to use technology
   C.2 No plan to use the technology
   C.3 No understanding of the technology by engineers

D. Technology exists outside the company or division.
   D.1 Insufficient time to use technology
   D.2 No plan to use the technology
   D.3 No understanding of technology by engineers

E. Technology does not exist
   E.1 Planned to develop technology
   E.2 Did not plan to develop technology

TABLE 5.4 Root Causes for Problem Generation

F. Inappropriate new product targets
   F.1 Inappropriate survey of user needs
   F.2 Inappropriate survey of competitive capability
   F.3 Inappropriate decision on user needs or competitive capability

G. Inappropriate design despite having technical design manual or design guidelines
   G.1 Error in spite of using manual or guidelines
   G.2 Designed by experience, that is, ignored design manual or guidelines

H. Technical manual or design guidelines did not exist
   H.1 Manual or guideline was planned
   H.2 Manual or guideline was not planned because of technology

An application of the T-type matrix. In Fig. 5.16, we show an actual application of a T-type matrix for a computer keyboard. This example comes to us from Hewlett-Packard Singapore. It was the first application by a young but enthusiastic design team, and they benefited tremendously from it.

The data in Fig. 5.16 was collected for seven subphases: from "Investigate & Define Specifications" up to "Manufacturing Release." In theory, we recommend that you collect data up to the point when customers use the product. In our example, this was not done mainly because the designers wanted to quickly analyze the data.

Note that in Fig. 5.16, the percentages of errors are also given in the lower portion of each box, in addition to the various sums. Look at the right-hand bottom of the matrix—at the Accumulated Sum box. Here we are reviewing the Problem Generation Phase. *You will notice that 95 percent of the errors occurred by the second subphase (the design phase). This data supports our statement that 50 to 80 percent of errors discovered in manufacturing and at the customer's site are generated in the early stages of design.* Therefore we reiterate: the project postmortem, including a T-type matrix, is an excellent tool to analyze and uncover the source of errors in new product design in order to improve the product development process.

Analysis of the T-matrix data. In Fig. 5.17, the data from Fig. 5.16 is analyzed after segmenting it into the categories listed in Tables 5.2, 5.3, and 5.4. Note that the specific lines from which to obtain the data are indicated in Fig. 5.15. In Fig. 5.17, the data is segmented into the following broad categories:

1. Reason for not discovering in the could-have-been-discovered phase
2. Lack of test technology
3. Root cause of problem generation

The errors in each of these categories from Fig. 5.16 were analyzed as to why they were not caught and then entered onto Fig. 5.17. Let us review an example in the "Reason for not discovering in the could-have-been-discovered phase." Look at Fig. 5.17 and read the product design components column for PCB (printed circuit board). We had a total of four problems that were not discovered. Of these, three occurred because of inappropriate checking criteria, and one occurred because checking was considered unnecessary. Now, go all the way down in the PCB column until you reach the "Root cause of problem generation" category. Three of these errors occurred despite having a technical manual; apparently the manual was not helpful. One of the errors occurred because the designer decided to use his experience instead of the manual. The corrective action will be to improve the manual and to request that designers use it.

Accum Non-D Sum	Non-D Sum	Accum Sum	Sub-Phase Sum	M. Release	Prod Proto	LAB Proto	Mold Prep	Model	Design	Investigate & Define Specifications	PHASE C: Problem Detection Phase / PHASE B: When Problem Could Have Been Discovered Phase / PHASE A: Problem Generation Phase	Investigate & Define Specifications	Design	Model	Mold Prep	LAB Proto	Prod Proto	M Release
8 / 14	8 / 14	8 / 12	3 / 12	1	3		1	3			Investigate & & Define Specifications	8						
42 / 74	34 / 60	44 / 67	36 / 55		2	5	22	5	2		Design		36					
42 / 74	0 / 0	44 / 67	0 / 0								Model							
42 / 74	0 / 0	47 / 71	3 / 5				3				Mold Prep		3					
57 / 100	15 / 26	64 / 97	17 / 26	1	14	2					Lab Proto		15		2			
57 / 100	0 / 0	66 / 100	2 / 3		2						Production Prototype		1			1		
											Manufacturing Release							
				2	21 / 35	7 / 11	26 / 35	8 / 12	2 / 3	0 / 0	Sub-Phase Sum	8 / 12	55 / 83	0 / 0	2 / 3	1 / 2		
				66 / 100	64 / 97	43 / 65	36 / 55	10 / 15	2 / 3	0 / 0	Accumulated Sum (of sub-phase sum)	8 / 12	63 / 95	63 / 95	65 / 98	66 / 100		
				2 / 4	19 / 33	5 / 9	23 / 40	8 / 14	0 / 0	0 / 0	Non-Diagonal Sum (of sub-phase sum)	0 / 0	19 / 86	0 / 0	2 / 9	1 / 5		
				57 / 100	55 / 96	36 / 63	31 / 54	8 / 14	0 / 0	0 / 0	Accumulated Non-Diagonal Sum (of sub-phase sum)	0 / 0	19 / 86	19 / 86	21 / 95	22 / 100		

NOTE: Number of Design changes / Percentage

Figure 5.16 Application of a T-type matrix to track design changes for a computer keyboard.

T-MATRIX APPLICATION: KEYBOARD PROJECT ANALYSIS

AREA	CAUSE CATEGORY	DETAILED CATEGORY	IC	PCB	MEM-BRANE	KEY-CAP	OUT-SERT	DOME-PAD	CASE PART	PAPER	DETAIL TOTAL	CAUSE TOTAL	TROUBLE TOTAL
REASON FOR NOT DISCOVERING IN COULD HAVE BEEN DISCOVERED PHASE	A – CHECKED	A-1 Checkers Mistake			1	1					2	16	57
		A-2 Inappropriate Criteria		3		1		1	9		14		
		A-3 Inappropriate Methods											
		A-4 Inappropriate Items											
	B – NOT CHECKED	B-1 Checking Judged as unnecessary			1			2	1		4	41	
		B-2 No Checking Item Existed	2			13	1		22		37		
		Component Total	2	4	1	15	3	1	32		57	57	
EVALUATION or TEST TECHNOLOGY ISSUES	C – EXISTED INSIDE	C-1 No Time to Apply				7		1			8	11	22
		C-2 No Plan to Apply					3				3		
	D – EXISTED OUTSIDE	D-1 No Time to Apply											
		D-2 No Plan to Apply											
	E – DIDN'T EXIST	E-1 Development Was Planned							2		2	11	
		E-2 Development Not Planned		2		7					9		
		Component Total		2		14	3	1	2		22	22	
ROOT CAUSE OF ERRORS OR DESIGN CHANGES	F – R&D TARGETS	F-1 Inappropriate Survey of Users Needs											66
		F-2 Inappropriate Survey of Competitors Capability											
		F-3 Inappropriate Decision on User Needs of Competitive Capability											
	G – TECHNICAL MANUAL EXISTED	G-1 Error in spite of using manual	3	3					15		21	24	
		G-2 Designed by experience instead of Manual		1		1	1				3		
	H – TECHNICAL MANUAL DIDN'T EXIST	H-1 Manual was Planned										42	
		H-2 Manual was not Planned				22	2	1	17		42		
		Component Total	3	4		23	3	1	32		66	66	

Figure 5.17 Analysis of T-matrix data.

From Fig. 5.17, the data has been converted into Pareto diagrams for each of the three categories. This is shown in Fig. 5.18. The Pareto diagrams display the weaknesses that were uncovered from the analysis. Based on the weaknesses, corrective action was taken in three areas:

- Improved inspection and checklist
- Introduced new test and evaluation technology and enforced the use of existing test and evaluation technology
- Improved technical manual with design guidelines for all designers

The information was conveyed to the entire design team, and training was provided where necessary. Subsequent analysis of the next keyboard design showed a marked reduction in design changes.

When and how to use the T-type matrix. Collecting, displaying, and analyzing the T-type matrix data takes some effort. The benefits, however, are tremendous. Here are some hints that make data collection easier for this important analysis.

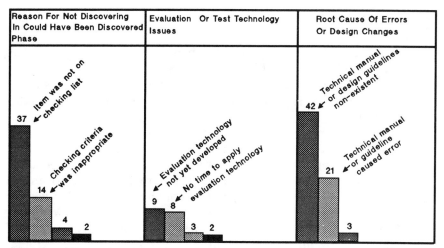

Figure 5.18 A Pareto diagram analysis of weaknesses in the design process (based on data from Fig. 5.17).

1. Provide a central record book for each project for recording every problem as soon as it is discovered. Where possible, direct entry into a computer database is preferable.
2. Convey the reasons and benefits of doing this analysis to one and all. Nobody should be reprimanded, instead R&D, marketing, and manufacturing staff members should understand that this is a method for ensuring continuous improvement in the new product development environment.
3. The matrix can be entered and computed in a computer spreadsheet such as Lotus 1-2-3.
4. It is difficult to commit to doing this analysis for all your new products, so try it on a few each year. The lessons learned in most cases will be generic and will steer you in the right direction. For example, they will raise the awareness for better design guidelines, ensure automatic applications of design guidelines via computer-aided design software, and improved testing and evaluation technology.
5. Use the T-type matrix within a project postmortem to get the maximum benefit.
6. The T-type matrix will be most useful when applied to the first-generation product in a series, for example, for your first keyboard, workstation, optical disk drive, or digital recorder.

Summary of project postmortem and T-type matrix

Approximately 50 to 80 percent of problems and design changes discovered in the later stages of manufacturing and at the customer's location are created in the early stages of the product development process. The reasons for this must be analyzed and understood and the problems reduced.

A project postmortem is crucial for understanding strengths and weaknesses in the new product development process. And within the project postmortem the T-type matrix will provide a very detailed analysis and understanding of design changes, problems, and errors. In fact, the project postmortem is a process that analyzes the effectiveness of all the processes in new product development.

Kaoru Ishikawa, the father of Japanese TQC, has this to say about analysis of new product development:[2]

> Whenever I am asked to help introduce a total quality control program to a company, I choose for my case study one of the company's new product development projects that is full of problems.

The lessons learned from the postmortem can be used to improve the entire product development process from collecting customer needs, forecasting, designing, and manufacturing to use by the customer. This continuous improvement of the product development process is the key to building products which are robust and successful.

Quality Assurance System

Every product manufacturing division, operation, or entity should have a quality assurance system. Such a system should include:

- Procedures for management of customer issues, feedback, complaints, and claims
- Standards for design, processes, purchasing, etc.
- Quality control procedures for the manufacturing process
- Overall coordination of the improvement process after customer and failure data is collected

We have discussed some of these items already, for example, the customer complaint and feedback system and design standards. We will not discuss items that have been mentioned, but we will show where they fit within the quality assurance system.

Look at Fig. 5.19 where we show the flow of quality assurance activities. The quality activities start immediately during design when the

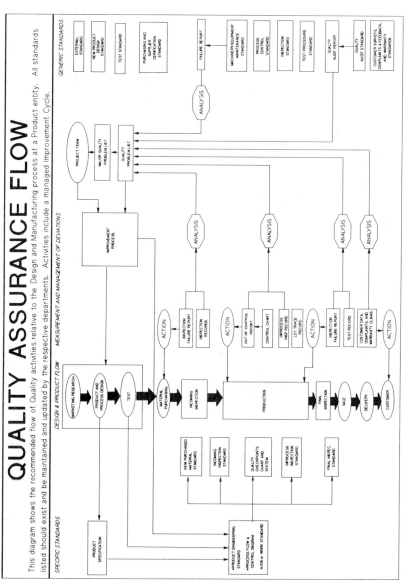

Figure 5.19 Quality assurance system.

product and process must be designed, and they continue until after product delivery to the customer. Figure 5.19 has the following highlights:

- Product specific standards that are listed on the left of the figure. These are required during the design of each product and include product specifications, process flow and control, and job or work standards. The job standards refer to the actual work procedures for each step in manufacturing. These items will then influence the following:
  - Specific standards for new, never purchased before, material and parts.
  - Incoming, in-process, and final inspection standards.
  - Quality checkpoints chart and systems. This will be discussed in great detail in the section "Quality Checkpoints Chart and Systems."
- Generic standards are listed on the right and include:
  - External standards, such as measurement standards.
  - New product design standards including design for reliability standards that we discussed earlier.
  - Machine and equipment maintenance standards.
  - Quality audit standards which are used to audit both the current manufacturing process and products on a routine basis—to ensure these confirm to the agreed upon standards.
  - Customer complaint and feedback system standards that we discussed in Chap. 2.
- On the immediate right of the design and product flow (with heavy black arrows), we have a series of activities that measure and manage deviation from standards. Examples are:
  - Inspection failures.
  - Customer complaints and claims after product shipment. We discussed this in detail in Chap. 2.
  - Control chart analysis and deviation report. We will discuss this in greater detail later in the chapter.

These deviations must be detected, analyzed, and eliminated. In addition, items for improvement are routinely identified.

Quality assurance improvement cycle

Look again at Fig. 5.19. The various deviations from standards require:
- Immediate short-term corrective action.
- Detailed analysis and understanding of the root causes.
- Generation of quality problem list, all items of which need resolution.
- Generation of a major quality problem list, items of which are given high priority and assigned to project teams for resolution.
- The resulting improvement cycle is used to influence marketing, product design, process design, and the current process.

This closed loop in the quality assurance system is very important. It runs in parallel to the product postmortem that we discussed earlier, except that this is an ongoing activity. The review of quality problems and the appointment of project teams to improve them should be managed by the quality department, working with the quality steering committee of the company.

Quality checkpoints chart and system

The quality checkpoints chart is a very crucial and important item for any manufacturing environment. It is not commonly applied. When applied, it is not always managed with sufficient discipline. That discipline is what makes the difference between a company making products with highly variable (or inconsistent) quality and a company with consistent quality.

Such a system is basic to all process planning and control. It provides a collation of all important and critical manufacturing data which should be documented and will include:

1. A production process flow chart including material status and its transformation into higher-level assemblies.
2. Reference to job standards and other documents that are necessary to complete each step in manufacturing.
3. Points in the process where measurement and control take place.
4. Responsible parties for managing the process and correcting out of control situations. This step will be supported by a formal documented out of control report.

Typically, the documentation is kept for each product that is manufactured. As manufacturing processes get simpler and more automated, the number of critical or control points will reduce. Nevertheless, the overall documentation and process must be maintained, managed, and improved by manufacturing personnel.

The process and quality checkpoint chart. In Table 5.5 we show a completed format for listing the process and quality checkpoints of a product during manufacturing. It is maintained on a Lotus 1-2-3 file or equivalent. Here is a brief explanation of each column in the table:

- *Material.* This refers to the specific part or assembly that is being used.
- *Flowchart step number.* There are two parts to this. *Incoming:* refers to incoming material or a subassembly flowing into the main assembly procedure. *Main:* refers to the main assembly steps. This can be merged into one flow. Refer to Table 5.5 for an actual example. The step numbers refer to the actual step in a separate flowchart of activities.
- *Process specifications.* This is the actual step of the work standard that describes how each step is done by workers or machine. This should be actual work procedures, not theoretical ones.
- *Critical control parameters.* Here we list the critical parameters that must be managed or controlled at each process step.
- *Frequency or sample size.* This explains how often the critical parameter is inspected.
- *Inspection or control method.* This explains the method of inspection of the critical parameter.
- *Chart used.* After the critical parameter is inspected or controlled, the data is entered and maintained in the document listed here.
- *Inspection item.* This is the actual item to be inspected. This is different from the critical parameter to be controlled. For example, the critical parameter could be dimension, while the inspection item could be gold bonding wire.
- *Specification limit.* This states the limit. When the limit is crossed, we have a failure, and corrective action must be taken. These limits are based on experience and data, for example, the upper and lower limits in a control chart. During normal day to day operation these limits would not be crossed. But when they are crossed we have an out of control situation that needs analysis and resolution.
- *Respond personnel.* This states the person who is responsible for checking the various critical parameters and raising the alarm when a specification limit is crossed.
- *Action personnel.* This states the person who must take corrective action when a specification limit is crossed. This activity includes an analysis of why the limits were crossed, resulting in an out of control situation, and the appropriate immediate and long-term corrective action.

Title: Process Assembly and Quality Checkpoints (Daily Management for Production Line)

Material	Flowchart In	Flowchart Main	Process Spec & job standard	No.	Critical control parameter	Frequency or sample size	Inspection or control method (Spec.)	Chart used	Inspection item	Specifications limit	Respond personnel	Action personnel
Silver-epoxy	1											
		2	Die attach 5956–5141–42	2.0	De shear strength	One bottle per shipment SS = 90 units	Die shear M/C	Record book		Min. 80 gram	E.O	IQA Eng.
				2.1	Storage in Refrigerator		—	—	—	—	U.O	Prod. sup.
				2.2	Kept at room temp. for 4 hr min. before opening bottle	—	Time and date	Record on bottle	Silver epoxy bottle	Min. 4 h	U.O	Prod sup.
				2.3	Change epoxy every 72 h	—	Change every night shift on Sunday & Wednesday	—	—	—	D.L.	Prod. sup.
Dice	3											
Leadframe	4											
		5	Die attach 5956–5128–42	5.1	DA insp. criteria	4 × 150 units per machine / shift	Visual @ 20 × min.	Trend chart	Die attach units	Acc. = 1, REJ. = 2	Inspector	Operator & tech.
						Inspect after major coversion/repair	Visual @ 20 × min.	Trend chart (write AC / AR at this inspection point)	Die attach units	Acc. = 1, REJ. = 2	Inspector	Operator & tech.
		6	Die attach cure 5956–5141–42	6.1	Oven temp.	Once a week	Temp. Profile	profile chart	IR oven	Soak time = minimum 6 mins.	E.O.	Engr.
						1 × shift	terminal	Recording book	IR oven	Zone 1:400 +/−5* C Zone 2:350 +/−5* C Zone 3:350 +/−5* C Zone 4:350 +/−5* C Zone 5:350 +/−5* C Zone 6:350 +/−5* C Zone 7:350 +/−5* C (Note * = Degrees)	D.L.	Tech sup.
				6.2	Speed	1 × shift	Terminal	Recording book		5 in / min	D.L.	Tech/sup.
				6.3	Die shear strength	5 units/sample 1 × shift	Die shear M/C	X-R chart	Shear strength	80 gram (min)	D/A insp.	Prod Eng.
Gold wire	7			7.1	Tensile test	3 spools per shipmt.	Tensile strength meter	Recording book	Tensile strength	Min. 8 gram Acc. = 0, REJ. = 1	IQA	IQA Eng.
				7.2	Visual	SS AQL = 2.5%	Visual 20 ×	Recording book	Gold wire	AQL = 2.5%	IQA	IQA eng.

Internal process doc. # : OPT-G-001   Product: Lamp   Revision: G   Date of issue: 15 May 1989   Page 1 of 3

Selecting quality checkpoints. Selecting the correct points for the quality checkpoints chart is crucial. How do we ensure that the correct points are selected? Here are some suggestions:

From experience. This is the most obvious method and relies on experienced production personnel. A documented list of critical production steps should be available for reference whenever a new product quality checkpoint chart is prepared.

Linkage to quality function deployment charts. If QFD has been used for product design, a lower-level QFD chart can be prepared to link the production process to crucial product characteristics. This provides a link to original customer needs. Refer to the discussion on the four phases of QFD earlier in the chapter.

Failure mode and effect analysis. After a detailed product flow has been completed, all production steps are listed. A FMEA can be done to highlight potential process problems. Refer to the discussion on FMEA earlier in the chapter.

Derived from customer issues. After a product is released to market, customers may give feedback on minor or major defects. This feedback is used to analyze the process and to determine areas that need better control. This information can be used to improve current process controls or to add new process control points.

The out of control report. When an out of control situation occurs, there must be a detailed analysis and corrective action must be taken—refer to Fig. 5.20 for an example. Specifically, such a report must address the following:

- *Description of the problem.* Describe what happened and where it happened; provide data.

- *Investigation of the causes.* That is, why the specification was not met or why the control limits were crossed. This requires an understanding of the root cause.

- *Immediate corrective action.* Here we list a quick fix or remedial action. This action will get things back to normal. This includes inspection for the defect or repair of the defect.

- *Preventive corrective action.* Here we list the preventive action. This will include elimination of the root cause; hence it will prevent recurrence of this problem. This is more important than the immediate action, but because of constraints the immediate action may be implemented first.

- *Checking effects of long-term measures.* This is to check on the effectiveness of the preventive measures and to ensure the root cause is eliminated.

File No. 89-23

		Control Chart Reg. No. B-05	
	Machine: Automatic Tester		
	Process: Assembly Test	Lot No.	
	Quality Characteristic: Fraction Defective at Test Station	Inspector:	
Description of Problems	Fraction Defective exceeded the upper control limit at the printed circuit test station for the computer mother board	P % ─ ─ ─ ─ ─ ─ ─ ─ ─ UCL = 2.1 ─ ─ ─ ─ ─ ─ ─ ─ ─ CL = 0.9 1 2 3 4 5 6 7 8 9	Date 7 December Time 3.50pm Discovered by
Investigation of Causes	The increase of failures was due to high failures of Integrated Circuit U39. Currently, there are 2 suppliers of U39. The circuit from Supplier A is giving an abnormal failure rate.		Date 8 December Name Engineer Date Name Date Name
Immediate Action	Only use circuits (U39) from Supplier B. If there is a shortage, then Supplier A Integrated Circuits will be pre-tested and faulty ones discarded – a special test fixture has been prepared for this purpose.		Action taken by Name Engineer Confirmed by Name Supervisor Date 8 December
Preventive Measures	Work with Supplier A to understand and estimate the problem. Increase purchase of Supplier B parts and stop all purchase from Supplier A, until Supplier A can eliminate problem.		Date 11 December Time 9.00am Confirmed by Purchasing Manager
Checking Effects of Preventive Measures	Supplier A has eliminated problem, which was traced to a process change. Data shows that Supplier B part has always been giving 40% lower failure than Supplier A. Hence, if Supplier A cannot reduce failure further, we may have to consider stopping purchases completely. Decision will be made in 6 weeks.		Date 28 December Name Engineer Confirmed by Purchasing Manager
	KEEP ON FILE FOR 5 YEARS	Department: A401 Computer System Division:	Updated Process Spec.#	Supervisor's Name	Eng. Mgr. Name

Figure 5.20 A completed out of control report.

In Fig. 5.20, we have given a completed example of an out of control report. The analysis and determination of corrective action are fairly straightforward. Note that the problem was traced to a process change at one of the suppliers; although the problem has been resolved, the supplier is in danger of being dropped.

Quality Assurance Sacrificed for Earlier Time to Market

Recently, even reputable Japanese manufacturers have been having a glitch of quality problems. *Business Week** reported that Japanese television makers Matsushita, Sony, Pioneer, and Toshiba recalled dozens of models that smoked or caused fires. In fact, a Pioneer TV set may have caused Japan's first major high-rise apartment fire, according to a government disclosure. Epson recalled about 100,000 laptop computers. Even highly reputable Toyota had to recall its luxury Lexus sedan to fix brake light and cruise control problems. Here is a list of recent recalls in Japan, compiled by *Business Week*.

Company/Product	Problem
Toyota Lexus automobiles†	Cruise control, brake light
Mitsubishi Motors Pajero Wagon	Accelerator
Isuzu Trucks	Loose bolt
Fuji Heavy Rex Minicars	Clutch contact
Yamaha Virago Motorcycles	Fuel leak
Honda Horizon Motorcycles	Transmission cog
Seiko Epson Laptop Computers	Circuit soldering
Toshiba TVs	High-voltage circuits
Pioneer Electric TVs	Circuit soldering
Matsushita TVs	Transformer insulation
Sony TVs	High-voltage circuits

Business Week questions if Japan is losing its quality edge. The answer is Not yet. The sudden surge of problems can be traced to the following reasons, compiled by *Business Week* and ourselves:

- Cheaper imported components from other Asian suppliers, who are not able to provide consistent quality.
- Lack of technical skill in new workers or, worse still, erosion of the work ethic.
- Lack of discipline in analyzing and correcting minor deviations.
- Introduction of high-technology parts and designs without proper understanding of how to control their quality or proper prediction of possible failure modes.
- The desire of shorter time to market, requiring a shorter design and manufacturing cycle. This may cause the elimination of certain tests or extensive tests which were normally done previously.

---

*Business Week, March 5, 1990.
†The only recall affecting the U.S. market.

The Sales Process

The sales process is often thought to be unstructured and not measurable, but it is really quite simple to structure the work into several activities and subprocesses. In Fig. 5.5b, we have shown several activities occurring in the marketing and sales environment—ranging from understanding market needs and opportunities, all the way to delivery and support of the product. All of the activities listed can be better understood and properly documented with performance measures, and training can be provided.

It would be foolish to suggest that a structured sales process will be a panacea for all sales problems—there is no substitute for an intelligent and creative sales representative. The techniques we will provide are those that many successful sales representatives or managers already use. The outcome of using these techniques can be a better managed and more predictable sales process, resulting in higher productivity.

In this discussion, we are not referring to retail sales but instead to items sold by a direct sales force—items such as instruments, computers, turbines, engines, machinery, and so on. We will not discuss everything shown in Fig. 5.5b, but instead we will focus on some of the less well-understood activities. I am grateful to Shailesh Naik, sales and quality manager at Hewlett-Packard Australia, for his help in editing this section.

The sales funnel

The concept of the sales funnel is well known to most marketing and sales people. This was popularized, in part, in the book *Strategic Selling*.[7] A sales representative doesn't just wait for orders, instead he or she must keep pouring prospects and suspects into a pipe or funnel and expect that some of them will come out through the other end as orders.

Items in the funnel can be segmented into those above the funnel, in the funnel, the best few, and orders. This is shown in Fig. 5.21. Each potential order must be tracked from inquiry, through the funnel, until it becomes an order. The steps that the potential order goes through will form the process.

We can now list the steps in the sales process. These steps will occur above, in, and at the bottom of the sales funnel, shown in Fig. 5.21. More specific steps of the sales process are listed in Table 5.6. This entire process should be well documented and training should be pro-

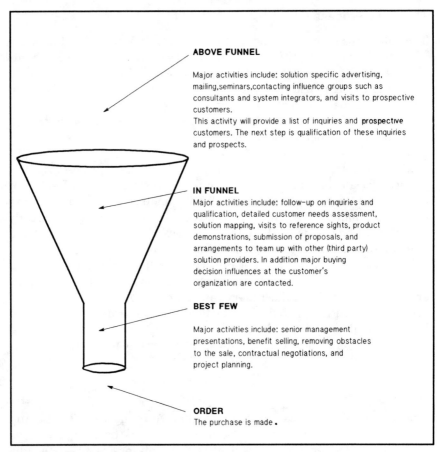

Figure 5.21 The sales funnel.

vided, which is listed in the very first step in Table 5.6. Improvements to the process should be made based on analysis from the sales won or lost postmortem and other analysis.

Based on our discussion, we can now come up with performance measures for the sales process, but before we do that, let's look at the sale situation from a different perspective.

TABLE 5.6 Steps in the Sales Process

Process step	Detail of the step
Sales planning	Provide training in the sales process, products, and solutions Business segmentation by products and markets Territory quota planning Customer contact plan Product introduction and demonstration to develop customer inquiries Marketing programs to develop customer inquiries

Selling	Customer contact and visit
	  Understand needs
	  Review competition
	  Check budget
	  Check product availability
	Qualification of customer
	  Meet buying influences
	  Suggest solutions
	  Understand intricacies and politics of situation
	  Review budget constraints
	Investigate
	  Study detailed needs
	  Provide solutions
	  Provide reference sites
	  Position competition
	Present detailed solution
	Close—get the order

Delivery	Deliver product
	  Keep commitments
	  Ensure trouble-free delivery
	Install product

Sales won or lost postmortem	Analyze sales won or lost
	  Understand success and failure factors
	  Understand competitive offerings
	  Improve current sales process

After-sales support	Provide after-sales support and service
	  Manage all accounts, ensure product and solution is meeting expectations; if not, understand why
	  Provide product support and service
	  Measure customer satisfaction
	  Collect and analyze all complaints and feedback

The sales situation

In Fig. 5.22, we show the relationship between sales forecasts and the actual results. Look at quadrant 1 in the figure. When a sale is forecast and won, we have success. What are the characteristics of such a situation?

- The needs of the customer were understood.
- Effective solutions were available for the customer.
- The sales activity and negotiations with the customer were well managed.
- If this success rate continues, sales productivity will be high.

In a similar fashion we can analyze the characteristics of the four quadrants in the sales situation. We show the characteristics of each quadrant in Fig. 5.23.

Now that we understand the characteristics, we can do a few things. First, it is obvious that as far as possible we want to be operating in quadrant 1—where a sale is forecast and won. Alas, this cannot always be the case. So, we have performance measures with appropriate targets for each of the quadrants. If we do that, we will be better able to control each of the four quadrants.

Identifying appropriate performance measures

We will now discuss some performance measures that can be used to manage the various subprocesses in sales. We have developed these measures from the sales process and the sales situation. The derivation of some of these measures is shown at the bottom of Fig. 5.23. More details on the performance measures are listed in Table 5.7 with targets and comments. The targets are typical of several sales managers we know. The performance measures and targets, however, will vary with your business and management style.

The performance measures in Table 5.7 should be measured monthly by comparing actual results against targets. Typically the data is collected from sales representatives via a daily or weekly activity sheet. This data is entered into a database, such as a Lotus 1-2-3. The performance measures can be plotted and compared to target. A deviation may indicate a need to intervene. For example, if the inquiry rate starts to decline, it may indicate weak marketing programs or a lack of new products

Daily Process Management

Forecast: \ Actual Results	Sale Won	Sale Lost
Forecasted Sale	Quadrant 1 Success	Quadrant 2 Failure
Non-forecasted Sale	Quadrant 3 A Lucky Win	Quadrant 4 Undeveloped Activity in seeking new customers

Figure 5.22 The four variables in the sales situation.

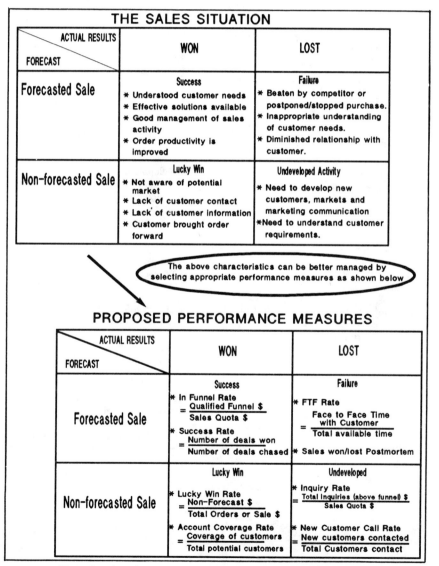

Figure 5.23 The sales situation.

TABLE 5.7 Performance Measures and Targets for the Sales Process

Performance measure	Typical target and limits (%)	Comments
1. Inquiry rate (3-month average) = (inquiry $/sales quota $)	700 (+200 to −200)	The inquiry or above funnel volume must be kept high, via marketing programs, seminars, product demonstrations, etc. The inquiries that get qualified will move into the funnel. *Note*: Sales quota $ is the planned sales volume.
2. In funnel rate (3-month average) = (qualified funnel $/ sales quota $)	300–400 (+50 to −50)	This measures the rate of qualified customer in dollars versus the sales quota in dollars. The target will vary, depending on the types of product or deal. Smaller-value products could have higher targets, while larger-value products could have lower targets. If this number drops, there is a possibility of not meeting sales quota in the future.
3. New customer call rate = (new customers called/ total customers called)	30 (+10 to −5)	This is important to ensure development of new accounts or markets by a sales representative or organization. Also, it is especially useful when new products are introduced.
4. Lucky win rate (3-month average) = (nonforecast $/ total order or sale $)	30 (+10 to −5)	These are unexpected orders, sometimes called *bluebirds*. If this rate gets too high, review the coverage of your business and market segment
5. Account coverage rate= (coverage of customers/ total potential customers)	40–100	Current and potential customers must be covered or visited. All customers who buy small-value products could be contacted by a telephone sales force. Customers who buy large-value products should be visited regularly, but 100 percent coverage may be difficult, hence a lower target. This can be better covered by requiring a sales representative to have a customer visit schedule.

Table 5-7 Performance Measured and Targets For the Sales Process (*Continued*)

Performance measure	Typical target and limits (%)	Comments
6. Success rate = (number of deals won/ number of deals chased)	25–35 (+10 to −5)	This measures a sales representative's or sales operation's productivity. The number could be smaller for low-value products but is expected to be higher for higher-value products (big deals) because fewer big deals will be chased, because more time is needed for a big deal. If this gets too low, review the qualification process and account coverage process (performance measure 5)
7. F.T.F rate = (face to face time spent with customers/ total available time)	40 (+10 to −5)	This is the time a sales representative or manager spends with customers. If it is low, there might be an administrative burden that requires attention. A possible alternative is the customer coverage rate, which includes customer face to face time and time spent working on customer needs. If this performance measure is used, the sales representative should ensure that time is spent with all potential customers, instead of a few.
8. Sales won or lost postmortem	Complete for all deals of value $100K or more	An understanding of why sales deals are won or lost is important. This will enable the sales force to improve its productivity by institutionalizing the lessons learned.

Sales won or lost postmortem

In the previous section one of the performance measures we discussed was a sales won or lost postmortem. The concept of a sales postmortem is similar to our previous discussion on a project postmortem. What we wish to do here is to understand why product or service sales deals are won or lost.

Objective and benefits of a sales won or lost postmortem. The objective is to better understand why sales are won or lost. The knowledge gained will be used to influence future sales in order to increase sales productivity and to improve the company's product development cycle. The sales won or lost postmortem will:

- Provide a better understanding of the competition
- Be able to better position your products or services against the competition
- Be able to conduct a better advertising campaign or better product seminars
- Provide information on better product or service pricing
- Provide feedback to designers on product or service weaknesses
- Provide data on resources needed to pursue future sales deals
- Provide information on how to improve training of sales representatives

The sales postmortem process. Next, we show a flowchart of the sales won or lost postmortem process. Note that the data is collected on a routine basis, but meetings and analysis are conducted after 3 to 12 months of data is available. The actual frequency will depend on the amount of data available, which is a function of the volume of sales. The postmortem team will be led by the specific product manager or marketing manager.

The process starts after each sale is won or lost. The data is then consolidated and analyzed. The steps are illustrated below:

1.0. Sales cycle completed and the sale is won or lost.
2.0. Sales representative collects pertinent data immediately. Data can be manually or computer tabulated and stored in central file.
3.0. Appoint postmortem team led by specific product manager or marketing manager.
4.0. Meetings planned:
    - Agenda
    - Data requested as per recommended format
5.0. Meetings take place (after 3 to 12 months of data is available):
    - Postmortem meetings held when sufficient amount of data is available
    - Data reviewed and analyzed

6.0. Issue report:
- Standard format
- Analysis and proposals

7.0. Review proposals for corrective action to improve current processes and practices:
- Final meeting with senior management
- Agreement on proposals
- Short- and long-term plans

Sales won or lost postmortem report format

1. *Objective:* Standard canned phrase, similar to the one given earlier in this section.
2. *Win or loss Pareto chart:*
   - Pareto chart of reasons sales were won.
   - Pareto chart of reasons sales were lost.
   - Pareto charts grouped by systems or specific products.
3. *Win or loss ratios:*
   - Win or loss ratios versus the trend over the years.
   - Win or loss ratios against each competitor.
4. *Competitive issues:*
   - Specific products, strategies, and solutions that we used that were successful against competitors.
   - Specific products, strategies, and solutions that competitors used that were successful against us.
5. *Analysis:*
   - List of strengths and weaknesses.
   - How can we leverage our strengths?
   - How can we reduce or eliminate our weaknesses?
   - Future product strategies for sales and factories.
   - Is there a third party or consultant that can help us do better?
   - How can we train our sales representatives better?

6. *Recommendations:*
   - Specific immediate action.
   - Specific long-term action.

Postmortem meeting guidelines. We provided guidelines for project postmortem meetings earlier in this chapter. Those guidelines are equally applicable for the sales won or lost postmortem meeting.

Collecting information on sales won or lost. There are several ways to collect this information on sales won or lost. We recommend the use of a formal data sheet for each sale won or lost. It will be preferable if the customer is visited after the event. When a sale is won, this will be easy, but when a sale is lost, it may be difficult. Hence, the visit needs to be positioned very carefully with the customer. The type of data to be collected will include the items shown in the data sheet in Fig. 5.24.

SALES WON/LOST DATA SHEET

* Name of Customer

* Situation: Product Type: _____   Value of Deal $ _____
               Primary Application _____
               Sales Representative/Manager _____

* Source of sales lead: ☐ Seminar   ☐ Trade Show
                       ☐ Telemarketing   ☐ Direct Mail
                       ☐ Repeat Sale   ☐ Others _____

* Particulars of 3rd Party or Consultant used in the deal

* Value of Service provided by 3rd Party/Consultant

* Sales Deal   ☐ Won against _____
            ☐ Lost to _____
            ☐ Value of Competitors winning deal _____
            ☐ Specific solution provided by competitor _____

* Reasons:
   • Customer's own words_____
   • 3rd Party/Consultant's own words _____
   • Sales representative's own words _____
   • Summary of reasons _____

* What could we have done better? _____

Figure 5.24 Sales won or lost data collection sheet.

The information needs to be collected routinely, by product. In a large organization with numerous products, you may wish to collect the information for products over a certain dollar value. When this data is collected over several deals, it will be possible to come up with a summary report which extracts the information from the data sheets and provides an analysis and recommendation. We show a portion of a postmortem report in Fig. 5.25a and b. Note the recommendations. These will be used to improve the sales process and future products.

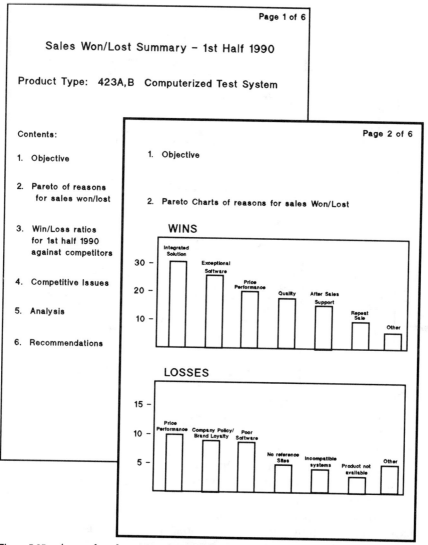

Figure 5.25a A sample sales won or lost postmortem report.

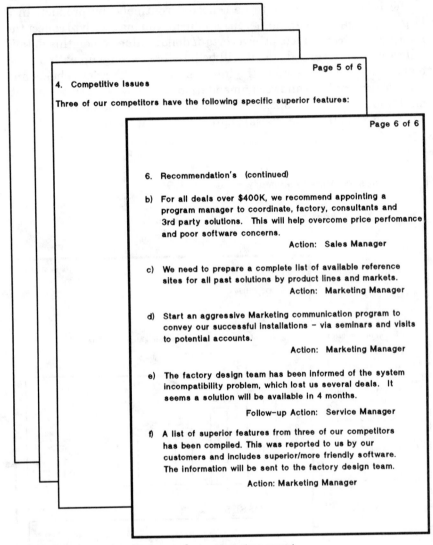

Figure 5.25b A sample sales won or lost postmortem report.

Summary of sales process

The sales process is often thought to be unstructured, but we have provided some ways to structure the process. We have proposed several performance measures with typical targets for managing the sales funnel. In addition, we have proposed a formal sales won or lost postmortem. These activities will help provide a better managed and more predictable sales process, resulting in a more competitive and productive sales force.

Managing Other Processes

We have discussed some of the key processes listed in Fig. 5.5a and b. The figure displays some of the processes in a manufacturing and sales environment—typically what you would expect in a consumer electronics, industrial products, or systems company. The specific processes in your company or organization may vary—you may have different products or services. But, the same concept of managing key processes will apply to your company or organization. Other key processes will include product delivery and software design. Let us briefly apply these concepts to the software design environment.

Key processes in the software design environment. Recently, there has been a lot of activity in improving software quality. Discussing process management in the software environment is beyond the scope of this book, but we offer the following observations based on the concepts discussed previously.

Currently, an important initiative in managing software processes is the software inspection program. These are reviewed by peers who use a formal process to examine the software design or development process for defects. Among the items inspected are the original software requirements—these original requirements are compared to the actual software product. The differences are registered as errors. For more information, we suggest you refer to the texts on software inspection that are available in the marketplace.

Based on our discussions in this chapter we recommend the use of the following methodologies during software design and development.

- *A better product definition process in software design.* A major source of errors in software design is not meeting the original requirements or customer needs. The use of QFD is recommended to capture customer needs and methodically convert them into specific design goals.

- *The use of reusable code.* This is similar to using design standards or reusable designs in new products to reduce design time and errors.

- *The project postmortem process.* This will be an extremely useful process to analyze errors and design changes in order to improve the software design and development process. We recommend using a process that is similar to the one discussed earlier in the section "Analysis of the New Product Development Process via the Project Postmortem and T-Type Matrix."

Reducing Design Time or Time to Market

In today's environment, reducing design time and time to market—that is, the time from product investigation to product delivery—is becoming increasingly important. To reduce time we offer the following suggestions:

- *Reduce time to do a key process.* Most of the key processes shown in Fig. 5.5 can be improved to provide higher speed and lower defects. This concept includes elimination of subprocesses or even of some processes.

- *Reduced design time via better software tools.* This is a very powerful method but beyond the scope of this text.

- *Project postmortems.* This improves your current design process by reducing design changes, forecasting errors, and delays. Postmortems and the subsequent corrective action help to institutionalize lessons learned. The result is an organization that learns.

- *Better product definition.* Use a better process, which includes QFD, to reduce missed customer needs, design changes, delays, and lost sales.

- *Design for reliability techniques.* This will reduce design errors and subsequent design changes that may cause further delay.

- *Reusable designs (both software and hardware).* This includes reuse of successful circuit designs, components, assemblies and software code. The use of these in a new product will save time.

- *Reduce manufacturing processes caused by numerous product options.* The fewer options available the better. This is not a proposal to reduce options available to the customer but instead is one to make many options into standard features as some car manufacturers do. By doing this you will be providing attractive quality and increasing sales. Refer to the discussion on the two dimensions of quality in Chap. 1.

- *Supplier certification and management program.* This will ensure purchase of approved products from approved suppliers. It will reduce the danger of problems during manufacturing that may cause shipment delays.

- *Computerization of activities.* Many of the above mentioned activities can be computerized, for example, QFD activity, FMEA activity, and postmortem data collection.

- *Just in time (JIT) manufacturing techniques.* This will increase manufacturing flexibility, so as to ensure rapid flow of products through the production line to the customer and to increase inventory turns (thereby reducing problems that come with high inventory). Information on JIT manufacturing is available in numerous texts.

- *Do some processes simultaneously.* Many of the processes listed in Fig. 5.5 can be done simultaneously instead of sequentially. Some of these are earlier supplier involvement for purchase of parts and earlier design manufacturing processes and preparation of process and quality checkpoints for new products.

Questions and Answers on Daily Process Management

Let's review some frequently asked questions on daily process management:

Q1. Should every process in an organization be documented and managed?

A1. No. It would be nice if you did that, but it is very difficult and could be counterproductive. You should focus on key processes—we have listed what we consider key processes in a design, manufacturing, and sales organization. In addition, every individual department in the organization must determine its own list of key processes—some of which will coincide with our master list.

Q2. For any given process, how do you ensure it is operating at an optimum level?

A2. There are several ways of looking at this. In the short term (1 to 2 years), you must ensure that the process operates within limits and targets. Any abnormalities and deviations must be immediately detected, analyzed, and eliminated. In the longer term (more than 2 years), there is no such thing as an optimum process. All processes must be improved and whenever possible subprocesses eliminated. This will come with continuous improvements, better designs, reengineering, and elimination of unnecessary activities (that were once thought optimum).

Q3. You discussed the quality assurance system and recommend the use of control charts and out of control reports. But this implies managing a process that is generating defects. I need to go beyond that—toward zero defects.

A3. You can, if your current level of defects is low. To move toward zero defects, consider the following data. Recently I visited a printer production line, with about 1/2 percent defects in the entire assembly line. The majority of the defects were caused by bad parts and workmanship. This leads me to recommend the following activities that should move you toward zero defects:

- Ensure all your purchased parts are good via an effective supplier management program.
- Eliminate workmanship errors by applying the concept of pokayoke, which we discussed in Chap. 4. This also implies automating as many production steps as possible. These steps should have a high process capability index to minimize or eliminate defects.
- Automate inspection wherever possible. Where you cannot, do 100 percent inspection—but this has risks.

Summary: Daily Process Management

We have discussed the importance of managing key processes. Well-managed processes can be leading indicators that predict good results for a company. Well-managed processes that are company standards will allow management to focus its energies in strategic areas.

We have identified several key processes in the design, manufacturing, and sales environment; and we recommend that you identify others in your organization. Managing your key processes and improving them will help increase efficiency, productivity, and quality.

These processes must be managed well on a day to day basis—hence the term *daily process management*. Daily process management means defining and monitoring a key process, ensuring it meets a target (within limits), discovering abnormalities, and preventing their recurrence. Success here means the ship (company) travels smoothly, while the captain (senior manager) focuses his or her effort on the destination (breakthrough objective).

The project and sales postmortem processes are very useful because they allow us to analyze the effectiveness of all our key processes in the design, manufacturing, and sales environments and to improve them. This will ensure that our organization learns from its mistakes. As we gradually decrease our mistakes, we will get more efficient and lower our costs and reduce design time and time to market. The results will be a more competitive and productive work force that provides increased customer satisfaction and higher profits.

References

1. Akao, Yoji. "History of Quality Function Deployment in Japan." From course notes of the Kaizen Institute of America, 1989, pp. V-A-1–13.
2. Akao, Yoji. *Quality Function Deployment.* Methuen, Mass.: GOAL, 1990.
3. Khushroo Shaihk Banu, the contributor of the QFD section, visited the California Cedar Company in Stockton, Calif., in April 1988. This company conducts research on pencils to help their customers. They verified that the house of quality on pencils was accurate.
4. Barker, Joel. *Discovering the Future, The Business of Paradigms.* Chart House, 1989.
5. "Does Success Have a Secret? Can LDC (Lesser Developed Countries) Compete in Free Market?" Stanford University, *Transactions PEASCON-87,* 2d Biennial Conference, Santa Clara, Calif., 1987, chap. 8, pp. 1–20.
6. Kano, Noriaki. "Problem Solving in New Product Development: Application Of T-Typed Matrix." *World Quality Congress Proceedings,* 1984, vol. 3, pp. 45–55.
7. Miller, Robert B., and Stephen E. Heiman. *Strategic Selling.* New York: Warner Books, 1986.

Chapter

6

Employee Participation

Quality is everybody's job.

A. V. FEIGENBAUM

Overview

An effective TQC effort will require the participation of every person in your company or organization. Never underestimate the worth of each individual. Consider the following news article (extracted from *The International QC Forum*, Nov. 1984):

> **If you think you are working hard, read on...A New Record of Suggestions—9310 Suggestions per Year**
>
> Mr. Koji Nakayama, QC Section at Utsunomiya Plant of Matsushita Electric TV Division, has established a record-high improvement suggestions reaching a total of 9310 suggestions in a year, all by himself. As to the secret of his inexhaustible source of suggestions, Mr. Nakayama says:
> "When an improvement idea hits me, I will go over it little by little on a daily basis. I try to put my idea in a presentable form and suggest it daily. I work on the idea on an average three hours every night after returning home. Complaints from my wife and kids? No problem—because I put myself at their disposal every weekend. So they don't bother me during weekdays. As to the source of hints, I'm a voracious reader and read all those newspapers, magazines, journal articles and watch TV programs that have something to do with my work directly or indirectly.
> "Good communication with other departments in the company is important to get richer information. Another source of suggestions is to develop (new) ideas for improvement after (each) improvement. I always keep a memo pad with me to jot down my ideas and observations."

This is just one of the ways that individuals can contribute to the success of a company or organization. We will discuss this and several other methods to harness the potential of every person. Our discussion will include the following:

- Quality circle, or team, activity
- Employee suggestion schemes
- Employee education
- Publicity promotion and recognition

Quality Circle, or Team, Activity

Numerous comparisons have been made between quality circles and quality teams. In general, quality circles have been considered as ongoing voluntary employee groups (operating at the lower levels of the organization) that solve problems in their work environment. Quality teams are considered ad hoc groups set up to solve specific quality problems. We think the distinction is fine, but in today's competitive environment it is difficult to accept voluntary participation.

Every employee needs to participate. Teams of well-trained employees on the production floor can solve process and product problems; teams of clerks in a bank can reduce paperwork or customer-related problems; management teams can solve service, product, or organization problems.

In many American and European companies the term *quality circle* has a stigma to it—in part this is due to initial false starts. Hence we suggest calling them anything you like—quality circles, quality teams, or project teams. For the remainder of this section, we will discuss how you can foster quality circle, or team, activity in your organization. In our context, the terms *quality circle* and *quality team* will mean the same thing.

Management of quality circle, or team, activity

A successful quality circle, or team, program is never an accident but the result of good management. Here are several key strategies that will help ensure a successful program:

1. *Organize a quality circle, or team, tactical committee.* This committee will best operate as a subset of the organization's quality steering committee, which we will discuss later. This committee will consist of senior managers with appropriate quality experts. At a department level it may simply consist of the department manager and his or her staff.

2. *Set expectations.* The committee should meet monthly to review and guide the effort. On a yearly basis—preferably at the beginning of the planning year—it should meet and set expectations and targets for the year, for example, as shown here:

- Participation rate of 80 percent of all employees.
- One project per year, per circle, or team.
- All managers must lead a management kaizan (MK) team. *Kaizan* means small or continuous improvement in Japanese.
- Review projects or areas for improvement. At small organizations or at department-level meetings, the specific projects should be reviewed, discussed, and approved.

3. *Identify and train facilitators.* All teams should get assistance from facilitators, except when there are experienced leaders. The facilitator is an experienced and well-trained individual. He or she can assist the team in staying on schedule, understanding the PDCA cycle, developing practical solutions, seeking technical help, and educating members.

4. *Provide education.* We have listed a suggested education program in the question and answer section. In addition, the following will be helpful:

- Maintain training data for all employees.
- Identify ongoing training needs. With time and progress, new training needs will become obvious. The facilitator or leader must identify these needs and arrange for training.
- As part of training, completed projects should be available and accessible as learning tools. This includes projects from within and outside the organization.

5. *Monitor progress.* There are several ways of doing this:

- Maintain a monthly status report of each circle, or team. Figure 6.1 gives a suggested format. The schedule shown in the figure should show planned and actual progress of each project. The goal should be for all projects to be completed within 1 year.
- Teams should select projects and commit to a schedule. Any deviations or slowdowns will require intervention and analysis by management. The facilitators will track and discuss issues and problems with the steering or tactical committee.
- Teams should be requested to keep minutes of all meetings and to track the steps of the PDCA cycle that they are following.
- Reward and recognize teams that are meeting regularly, have good attendance, or are exceeding agreed upon schedules. This will require regular feedback of activity, probably by the facilitators. Rewards can be tokens of appreciation.

Circle Name	Leader	Facilitator	Theme Project	Monthly Status											
				J	F	M	A	M	J	J	A	S	O	N	D
1. Startrek	Able	Rogers	Reduce Scrap	A	B	B	C	D	D	D	F	G			
2. Etc.															

QUALITY CIRCLE/TEAM, STATUS REPORT DEPARTMENT: _____

Key A = Projected Selected
 B = Plan Stage
 C = Do Stage
 D = Check Stage
 F = Act State + Documentation
 G = Presentation

Figure 6.1 Quality circle, or team, status report.

6. *Promote, recognize, and reward.* Ongoing promotion, recognition, and rewards are a must. Promotion will include the following activities:

 - Display team members' photos and project data on bulletin boards. Completed projects can be displayed and published in monthly newsletters.
 - Regular quality circle, or team, conventions are a must. Departments should hold their own conventions where circles, or teams, can present their projects. The best department projects can be presented at the organization's annual quality convention. A large organization can hold more than one convention a year.

- A few outstanding teams from the annual convention can go on to represent the organization at state, country, or international conventions. This not only provides recognition of and esteem for the best teams, it also exposes them to activities and new techniques outside the organization. This can be highly motivating.
- At annual quality conventions, the best projects should be graded and selected by a team of judges, comprising experienced managers. A recommended grading criteria is given in Fig. 6.2. We recommend

EVALUATION & SCORING SYSTEM FOR QUALITY CIRCLES/TEAMS

CIRCLE/TEAM: _____

	EVALUATION CRITERIA	MAX.	ACTUAL
CIRCLE PROCEEDING (10 PTS)	1. Meetings are held regularly and members attendance has been good.	2	_____
	2. Ideas contributed by members use of and improved upon.	2	_____
	3. Difficulties experienced and overcome during team's proceedings.	3	_____
	4. Help sought from other people, sections or departments inside and outside the organisation when required.	3	_____
PROJECT SELECTION (10 PTS)	5. Project selected based on study of background information constraint and past data.	5	_____
	6. Project meets the needs of the people, sections or departments served by the team, i.e., the customers – or – 7. Selected from department objectives or Annual Hoshin plan.	5	_____
PROBLEM DEFINITION (5 PTS)	8. Problems clearly identified and defined.	3	_____
	9. Target explained.	2	_____
ANALYSIS TECHNIQUES (20 PTS)	10. Techniques and methods such as fishbone chart, pareto diagram, graphs, check sheet, etc. effectively used.	10	_____
	11. Systematic approach to identifying and verifying the most probable causes. (Team used PDCA approach correctly).	10	_____

page 1 of 2

Figure 6.2a

EVALUATION & SCORING SYSTEM FOR QUALITY CIRCLES/TEAMS			
CIRCLE/TEAM: _____			
	EVALUATION CRITERIA	MAX.	ACTUAL
CORRECTIVE ACTION AND IMPLEMENTATION (15 PTS)	12. Alternative solutions considered. 13. Solutions are properly evaluated. 14. Recommended solution(s) is/are sound and practical. 15. Implementation was effective.	2 3 5 5	___ ___ ___ ___
RESULTS ACHIEVED (15 PTS)	16. Tangible results achieved. 17. Intangible results achieved. 18. Variation(s) between results & original target(s) is/are explained.	5 5 5	___ ___ ___
STANDARDIZATION (10 PTS)	19. Standardization carried out by changing procedures or through other arrangements. 20. Follow up action taken to ensure that new procedures are maintained.	5 5	___ ___
SELF-EXAMINATION AND FUTURE PLANS (5 PTS)	21. Team's next project stated and reasons given. 22. Team is aware of its limitations and potential problems. 23. Team gives proposals to overcome them.	2 2 1	___ ___ ___
PRESENTATION (10 PTS)	24. Team presents in an interesting manner. 25. Presentation is well organised. 26. Involvement by members in the presentation. 27. Audio visual aids used effectively with important points highlighted and clearly explained. 28. Presentation easily understood by listeners.	2 2 2 2 2	___ ___ ___ ___ ___
	TOTAL POINTS (100 MAXIMUM)		☐

Note: The points awarded range from 1 to 10 depending on the importance of the criteria.

Figure 6.2b

use of this criteria. The judges should be experienced in quality circle, or team, techniques. Also, consider getting external, experienced, judges.

- Separate MK team conventions should also be held. Consider managing the MK activity with the same rigor as regular quality circle, or team, activity.

An important afterword on quality circles and teams

When the concept of quality circles was first introduced, participation was voluntary. In today's highly competitive environment, everybody must chip in. Participation is no longer voluntary—it is essential for success. Hence, quality circles, or teams, must be encouraged at all levels in the organization; this includes formation of management or cross-functional teams.

These teams will work on different types of problems. The lower-level teams will resolve problems or improve processes in their own group environment. The management teams will resolve problems within their departments or across departments or divisions. In many cases, management projects will be items selected from the department or organization's annual Hoshin plan. The process used will be similar—the PDCA improvement cycle. To achieve high levels of participation, training of all employees and management commitment is essential.

Employee Suggestion Schemes

In the overview to this chapter, we quoted a very prolific person who contributes over 9000 suggestions per year. Certainly not every employee can do this, but the potential is enormous. Many American and Japanese companies have suggestion programs.

In Japan, many companies obtain an average of more than 20 suggestions per employee per year and up to 90 percent are implemented. The results can be millions of dollars of savings and better-quality processes, products, and services. In 1987, about 350 large Japanese companies are estimated to have saved about $2 billion from employee suggestions.*

In the United States, IBM has a successful employee suggestion program in which the contributor gets a percentage of the savings. In Germany, the Siemens Company also has a very successful suggestion scheme. The basis for a suggestion program is to allow all employees to have a say in improving things that they think are wrong, to allow bright ideas to be captured formally, and to recognize that the company's senior management does not have all the answers.

---

*Data on suggestions schemes comes from a Japan Human Relations Association (JHRA) survey.

Guidelines for an employee suggestion scheme

Here are some guidelines for an employee suggestion scheme. The bulk of the guidelines come courtesy of Hewlett-Packard Singapore.

- *Objectives of the employee suggestion scheme*
  - To support the participative management philosophy of the company.
  - To allow employees to give logical and practical suggestions for improvement in their work environment regarding safety, processes, products, services, systems, quality, design, etc.
- *Specific areas applicable for suggestion schemes.*
  - Production materials and its flow.
  - Design and layout of production floor.
  - All systems and processes.
  - Design of equipment, tools, and fixtures.
  - Work environment.
  - Safety-related issues.
  - Quality and design of products and services.
  - Work procedures.
  - Information flow.
  - Customer services and customer relationships.
- *Specific areas not applicable for suggestion schemes.*
  - Personnel policies and guidelines.
  - Salary and wage administration.
  - Personal grievances.
  - Human conflicts.
  - Items within the direct job responsibility and assignment of the employee, that is, what the employee is supposed to do.
- *Rules and regulations.*
  - Suggestions should be submitted on a standard form. An example is provided in Fig. 6.3.
  - All concerns must be accompanied by a suggestion before they can qualify for assessment. Proposers should approach their immediate supervisors for assistance wherever necessary.
  - The judges' decision is final.

| BRIEF DESCRIPTION: _____ | REGISTRATION NO.: _____ |

[hp]   **EMPLOYEE SUGGESTION SCHEME**   **TQC**

Name: _____
Employee No.: _____
Department: _____
Location Code: _____
Date of submission: _____
Supr-in-charge: _____
QC CIRCLE PROJECT: ☐ Yes ☐ NO

PHOTO
(OPTIONAL)

Guidelines:
A. Briefly describe present condition, method or practice.
B. Details of your suggestion for improvement.
C. Please write neatly or type.
D. Use additional sheets if neccessary.
E. For filing and tracking purposes, please give a brief description of your suggestion (not more than 10 words) at the top left corner.

PRESENT CONDITION:

_____
_____
_____
_____
_____
_____

YOUR SUGGESTION:

_____
_____
_____
_____
_____

Is your suggestion already implemented?
☐ Yes ☐ No
If yes, date of implementation:- _____

MANAGER's COMMENTS:

For implementation : Yes/No
Responsible person : _____
Date of completion : _____
Net Savings
(First Year only) : S$ _____
Type of Prize

Figure 6.3a

- The company reserves the right to make changes to the suggestion scheme and its reward system whenever necessary.
- Points obtained in a suggestion scheme grading system (see next item) may or may not be accumulated. The decision to accumulate will depend on the types and amount of rewards given.

to be completed by assessor
(For individual or group suggestion only)

S$/year

(1) No. of labour units or man-hour saved x labour rate (S$ /unit labour/mth) X12 = ☐

(2) Man-hour reduction (hour/unit x no. units produced/mth x hourly rate (S$ /hr) X12 = ☐

(3) Cost Savings/unit x no. of units produced/mth X12 = ☐

(4) Other savings/mth: ● Occupancy X12 = ☐

　　　　　　　　　　● Auto-expenses X12 = ☐

　　　　　　　　　　● Operating supplies X12 = ☐

　　　　　　　　　　● Others X12 = ☐

　　　　　　　　　　　　　　　A Total Savings = ☐

(5) Capital Investment x Depreciation rate/mth X12 = ☐

(6) Other incremental cost/mth: ● Labour X12 = ☐

　　　　　　　　　　　　　　● Materials X12 = ☐

　　　　　　　　　　　　　　● Others X12 = ☐

　　　　　　　　　　　　　　　B Total Savings = ☐

A Total Savings/yr — B Total Costs/yr = C Total Net Savings/yr

Criteria	Measures	Grades			Score 1st Assessment	Score 2nd Assessment
	Possibility of immediate Implementation	Cannot or has been implemented 0 Point		Will be implemented ☐ can be implemented ☐ 10 Points		
Idea	Degree of originality	Negligible 0 Point	Some 8 Points	Significant 15 Points		
Effort	Amount of effort in generating the suggestion	Negligible 3 Points	Some 8 Points	Significant 15 Points		
Customer Satisfaction	Customer Satisfaction Level	Negligible 0 Point	Some 5 Points	Significant 15 Points		
Net Savings	First year net Savings as a result of implementation	Zero — 0 point 2 points for each nearest $1000				
Safety	Degree of safety improvement	Negligible 0 Point	Some 5 Points	Significant 15 Points		
Quality	Degree of quality improvement	Negligible 0 Point	Some 5 Points	Significant 15 Points		
Others	Additional points could be given for other criteria not mentioned above. In this case give reasons: (maximum of 15 Points)					

Points to be awarded if suggestion will be or already implemented.

TOTAL NO. OF POINTS

Reward System
==============
Total score	0-14	15-49	50-79	>80
Prize value	Nil	$10.00	$40.00	$80.00
Certificate	Thank You	Bronze	Silver	Gold

Date of final Assessment: _____

Figure 6.3b

- *A proposed grading system.* Refer to the form in Fig. 6.3. All individual and group suggestions will be graded into thank-you, bronze, silver, or gold awards based on a point system. The points given will depend on the following criteria. (The assessors are not required to complete the cost savings computation for suggestions which will be graded thank-you.)

- Idea
- Effort
- Customer satisfaction
- Net savings
- Safety
- Quality
- Others

- *Reward systems.* Rewards systems can vary. The system shown here comes from the Hewlett-Packard Singapore suggestion scheme. We also mention some other reward systems.
  - For individual and group suggestion only, Table 6.1 will apply. This gives the reward versus score achieved in the suggestion.
  - For a group suggestion, the prize has to be shared among the group members.
  - For every five suggestions (from thank you to gold award) submitted within each fiscal year, the originator will receive an additional reward of $20.
  - Each year, the best six gold awards will get a special Best suggestion of the year award—typically a plaque and token of appreciation, such as a designer watch.
  - Each department must submit its claim voucher to the administration department at the end of each month and collect the cash awards.

Other reward systems can have features such as:

- Gifts, tickets, dinner vouchers, etc.
- Accumulation of points during each year. At the end of each year, the total points accumulated can be used to collect awards.

TABLE 6.1 Reward versus Total Score

Total score	0–14	15–49	50–79	80 & above
Prize value	nil	$20	$100	$200
Award given	Thank-you	Bronze	Silver	Gold

- A percentage of the dollar savings up to a maximum dollar value of, say, $25,000. IBM uses such a system.

- *Guidelines for all supervisors and managers*
    - The immediate supervisor must decide whether the suggestion raised by the proposer is within the job responsibility or assignment. Suggestions that are within the direct control of the proposer will not qualify as a suggestion.
    - Response time to suggestion after submission varies. For a suggestion requiring the proposer's own department to assess it, it is 1 week. For a suggestion requiring another department to assess it, it is 2 weeks.
    - If the actual assessment time exceeds the guideline, the responsible supervisor must explain the reasons for the delay.
    - The immediate supervisor can approve suggestions deserving a $20 prize. The department manager can approve up to a $100 prize. All $200 prizes must be approved by the functional manager.
    - Under the "net saving" assessment criterion, only the first-year savings will be considered.
    - The immediate supervisor is responsible for assisting the proposer in proposing a suggestion to resolve logical concerns which are beyond his or her capability.
    - The immediate supervisor should reject all frivolous suggestions or suggestions outside the company's control, for example, "build a bridge across a nearby river in order to reduce time taken to reach the company." This will save time and effort.

Promotion and measurement of activity

Constant promotion and encouragement will be necessary to make the suggestion scheme successful. In a separate box, we have given a random sample of the type of suggestions you can expect if you start a suggestion scheme. Here we will mention how you can promote and measure success.

It will be useful to have department targets for the number of suggestions desired and constant encouragement by managers. Promotion should focus on:

- Productivity gains not cash gains of employee
- Supervisor-worker meetings in order to solicit and encourage contributions
- Giving awards in public at department or company meetings

Measurement of success can be done by tracking the suggestion scheme activity. The following performance measures can be used. They are given in order of implementation, that is, what you will measure initially and what you should measure as you get more experienced and sophisticated.

- *Initial phase*
  - Number of suggestions per employee per year.
  - Percentage of employees participating.
  - Time to respond to contributor.
  - Time to implement suggestion.
- *Mature phase*
  - Yearly goals for suggestions per employee.
  - Quality of suggestions. One measure for this is percentage of suggestions implemented.
  - Cash savings per year.

Flow chart of suggestion scheme tracking system. For the suggestion scheme form, shown earlier, we provide a flowchart of activity. This is shown in Fig. 6.4. Note that suggestions with higher scores go progressively to higher levels of management. This ensures checks and balances, prevents abuse of the system, and allows senior management to be involved. Also included are a list of suggested performance measures.

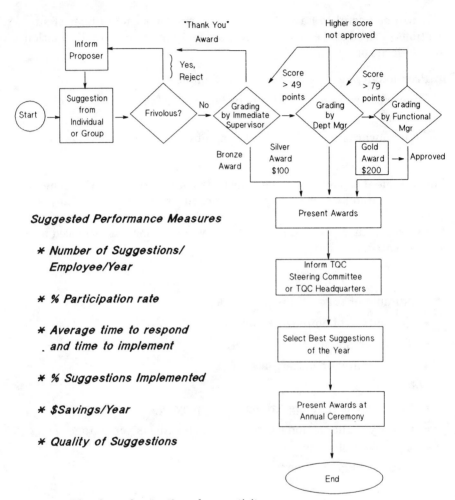

Figure 6.4 Flowchart of suggestion scheme activity.

> **What Type of Suggestions Can You Expect?**
>
> Suggestions will range from frivolous to excellent. With time and employee education, the quality and number of suggestions will improve. Here is a list of randomly selected suggestions that were taken from Hewlett-Packard Singapore and Hewlett-Packard Japan:
>
> - Sell, in order to recycle, waste material instead of discarding it. Result = savings $30,000 per year.
> - Eliminate misinsertion of connectors, which causes 24 percent rejected connectors, by using a keyed connector. Result = dollar savings were marginal, but rejects reduced to 0 percent.
> - Modification of an assembly work-holder to increase production speed. Result = savings of $12,000 per year.
> - Improvement of copper plating process. Result = savings of $16,000 per year.
> - Propose a new application for a Hewlett-Packard distance measuring instrument. Result = increased sales of $550,000 per year.
> - Reduced number of copies of account payables invoices. Result = savings of $5000 per year and reduced paperwork burden imposed on some managers.
>
> Altogether, this is a very impressive list. It represents improvements above and beyond what an organization's management would even think of doing. It helps enormously, then, if every member of the organization keeps his or her eyes and ears open and suggests ways to improve products and processes resulting in increasing productivity, profits, and customer satisfaction.

Education in Quality Methodologies

Education in quality methodologies is crucial and necessary if a company is to succeed and achieve its quality vision, to increase its productivity, and to ensure increasing customer satisfaction. Listed here are some basic quality methodologies that should be part of an employee education effort.

Course and contents	Target audience
1. *Introduction to TQC.* Current trends and techniques of quality; the why, what, and how of productivity quality and customer satisfaction.	All employees (special modification for senior managers, with a business and profit focus, is recommended).
2. *The PDCA improvement cycle.*	All employees.
3. *Seven quality control tools.* Refer to App. 2. Ensure that overlap between this and quality circle training is minimized.	All employees.

(Continued)

Course and contents	Target audience
4. *Design of experiments and Taguchi methods.*	All engineering staff.
5. *Quality circle member training.* Graphs, data collection and checksheets, Pareto diagrams, brainstorming, cause and effect diagrams, the PDCA cycle.	All nonmanagerial employees.
6. *Quality circle leader and facilitator training*, plus techniques for working in groups and communication and presentation skills.	All supervisors and managers
7. *Design for reliability techniques.* For example, training in thermal design component derating, and FMEA. Refer to the discussion in Chap. 5. Other training as appropriate with the products you may manufacture.	All engineering staff.
8. *Quality function deployment (QFD) training.* Refer to Chap. 5 for more details.	Marketing, R&D, and other engineering staff. Management must understand concepts.
9. *Sampling techniques.* Types of sampling and benefits of sampling.	All professionals and managers.
10. *Survey techniques.*	All professionals and managers.
11. *Seven new quality control tools.*	All staff but only after the seven quality control tools are well understood and used frequently.
12. *Hoshin planning.* Refer to the entire discussion in Chap. 3.	All professionals and managers.
13. *Customer satisfaction.* What is customer satisfaction? How to achieve it? This will include a session on the customer complaint and management system, discussed in Chap. 2.	All staff.

Publicity, Promotion, and Recognition

Ongoing publicity, promotion, and recognition of the quality effort is extremely important. Here are some suggestions to get you on the right track:

- *Monthly quality newsletter.* This can give recognition to successful quality circle projects and good suggestions and can convey new initiatives, ideas, and techniques.

- *Display boards.* Display of successful projects, quality circle members and their projects, all accepted suggestions, plans for the year, and new ideas. Displays must be managed and updated regularly; otherwise employees recognize quickly that management is not serious.
- *Recognition.* Provide recognition for employees who have contributed to the quality effort. This can be done in the newsletter, display boards, and other activities such as:
  - Departmental and company quality conventions
  - Sending the best quality circles to country and international quality conventions
  - Rewards for the best suggestions of the year

Some Questions and Answers on Employee Participation

Here are some answers to some recurring questions on employee participation:

Q1. Tell us more about a quality circle, or team.
A1. The answer to this question has several facets: The quality circle would come from the same work area for a lower-level team, for example, from production workers, technicians, engineers, material handlers, clerical workers, or waiters. This group would be led by their supervisor or by a fellow employee. A facilitator may be needed to help them with the improvement process.

The quality circle, or team, consisting of higher-level workers, professionals, or managers can have members from different work areas, for example, employees or managers that form teams from different departments or different divisions or between a company and its suppliers. The team would be led by an experienced manager. A quality circle, or team, can also be a group from the same area but consisting of a senior manager and his or her staff.

Q2. What are the objectives of quality circles, or teams?
A2. To improve the company or organization's performance. This includes increasing customer satisfaction, reducing product and service costs, and increasing yields in manufacturing and service processes.

Q3. What type of education is provided to quality circles, or teams?
A3. At the outset, the following education is needed:
- Understanding and use of the seven quality control tools, specifi-

cally, graphs, data collection and checksheets, and Pareto diagrams.
- Brainstorming.
- Cause and effect diagrams.
- The PDCA improvement cycle. This will be the cornerstone of the training for a quality circle.
- Working in group techniques—for team leaders and facilitators.
- Communication and presentation skills for all.
- As the circle, or team, matures, the other seven tools can be taught. Refer to App. 2 for details on the seven quality control tools.
- Eventually, with greater maturity, the seven new management tools must be taught.

Q4. Why do some quality circles fail?

A4. All too often, a well-intended effort to start quality circles fizzles out. Listed below are some of the causes.

- Insufficient education or training. The right tools and techniques have to be provided before a start can be made.
- Overeducation can cause problems. For example, there is no need to teach all seven quality control tools. The list shown above, in the previous question, is a good starting point.
- Lack of trained facilitators to help quality circles in the problem-solving process.
- Quality circles are often made up of people from different work groups. Although possible, this is not advisable in the early stages of growth. These lower-level circles consisting of production workers, clerks, etc., should consist of members of the same work area. Only more mature circles, or teams, for example, engineers and managers, should attempt to form groups from different work areas.
- Solving problems outside the control of their own work area or process. This is one of the most common problems for failure. When quality circles are formed, they should confine their initial activity to their own work area or process.
- Lack of time to meet. Management must allocate time for circles to meet—in slack times and during high-growth periods.
- High expectations by top management. Expecting quick results or

large savings is quite common. When this doesn't happen, it leads to disappointment and often withdrawal of support to quality circle activity. Typically, a circle may complete one project a year. If two projects are completed a year, it is considered very good. It must be emphasized that quality circles are not a panacea for a company's problem. We estimate that only 20 percent of all quality problems are solved by quality circles; the remaining 80 percent are attributable to and controllable by management.

- Not accepting a quality circle's proposal can have negative effects—it is the best way to kill a quality circle effort. This can be avoided at the outset by confining a circle's activity to its own work area or process. Further, a well-trained leader and facilitator will ensure that the team solves the right problems and provides good solutions. Management acceptance to the circle's proposal then becomes a fait accompli.

- Lack of exposure to quality circle techniques. Circle members must get exposed to activity outside their work area—other companies, even other countries. Regular quality circle conventions are a popular means to foster such exposure.

- Lack of management support is probably the major reason why quality circles fail. The reasons mentioned above are all controllable by management. Therefore, management must provide the resources and encouragement for a quality circle effort to succeed.

Q5. Why do quality circles succeed?

A5. There are numerous companies that have a high degree of participation—over 90 percent of all employees—in quality circles and management quality teams. The basic ingredient is management commitment, support, and encouragement. We list the main reasons here:

- Give people, especially from lower-level circles, time off to meet. This is usually 2 hours a week.

- If a meeting takes place outside office hours, always entertain overtime claims. After all, employees are working to solve company problems.

- Hold regular in-house quality circle conventions—one or two a year to encourage participation, share successes, and to offer recognition. All managers should attend these 1-day presentations.

- Consider a reward scheme for quality circles for each project completed. This can be a small but significant token of recognition for their efforts.

- Each year, consider sending the best circle on a working holiday to visit and present in other parts of the organization, often in other countries.
- Education in quality circle techniques must be made mandatory for all managers, engineers, supervisors, and employees. Quality circle training must be a core training course in the company.
- Set the expectation that all employees must participate in quality circles. Their achievements can be discussed during their annual performance evaluation. This reinforces their belief that management is totally committed to making quality circles successful.
- Set high expectations in quality improvement for all departments. High-quality products and services must be the goal. Target for quality improvement every year. Get employees to understand that customers expect top quality and high satisfaction; therefore, everyone has to work to meet this requirement—through quality circles or other ways.
- Managers should occasionally attend quality circle meetings. Managers should also review a formal presentation of each circle's completed project in order to endorse findings and recommendations.
- Higher-level quality teams—MK teams—must be formed in every company. This allows senior management to participate in improvement projects and to set an example for the rest of the organization to emulate. Typically, these management teams will not participate in quality circle conventions or get rewards. However, separate MK conventions can be held where managers can discuss and exchange ideas. Every manager should have a personal improvement project or a department improvement project or should participate in a cross-functional task force. Refer to the discussion on cross-functional task forces in Chap. 3. All projects must be linked to a Hoshin or Daily Management plan.
- Finally, management must exhibit lots of patience. Rome was not built in a day; neither can quality circle success. Managers have to support quality circle activities through good times and bad, through periods of intense activity and low activity, through crises and calm. And then, success will be theirs.

Q6. Suggestion schemes work well in the Orient but our culture makes it impossible to introduce such a scheme here.

A6. On a recent visit to Germany, we heard this comment from a senior German manager. On further investigation, we found that Siemens—a very German company—has had a very successful employee suggestion scheme for over a hundred years. So much for the comment that German culture restricts suggestion schemes. Hence, such comments are typically a smoke screen to avoid doing something. Suggestion schemes will work in any culture or company—but they require constant promotion and encouragement and an atmosphere where lower-level employees have no reason for fear. In addition, the quality department or a TQC headquarters function must facilitate this activity and help overcome roadblocks.

Q7. How do we ensure we have a successful quality circle, or team, program?

A7. We have already conducted a separate question and answer session for quality circles and given you a detailed list of suggestions in this chapter; but a few more comments are in order. Quality circles, or teams, will seldom survive in a vacuum, that is, when there is no management support. Quality circle, or team, activity must be part of an overall companywide quality or TQC effort. This way, management will commit to it, provide resources, and manage the activity, and employees will know that management is serious about the whole thing.

Q8. Our employees are not trainable.

A8. Then you have a problem and your company may fail in the marketplace. Your employees are your most valuable resource; if they lack basic education, give it to them and then train them in quality methodologies.

Q9. My priority is to get products out of the door; I will be diluting my resources if I start to promote quality circles and suggestion schemes.

A9. If you run a small operation—say less than 100 employees—or if you are managing a start-up operation, your concern is justifiable. But if you run a larger or mature operation, you need to reconsider. Employees today want to participate and contribute—the larger the organization, the greater the need. As we said earlier, there are only so many things that management can focus on, but employees can devote their ideas and energy to improving processes and products that they are close to. Also, unfortunately, management does not have all the answers—getting employees to contribute via the various systems suggested can only help, not hinder. The outcome will be management and employees working in unison toward a common goal of higher productivity, increased customer satisfaction, and of course, increased employee satisfaction.

Q10. Should the quality or TQC department conduct training?

A10. Preferably not. All training should be managed by a central training department, such as the human resources or personnel department. If you conduct quality training via the quality department, you have a situation where quality is viewed as a separate program—often to be ignored. The quality department can help design the quality training program, but it must be managed by one central training department and, preferably, conducted by experienced employees and managers.

Summary: Employee Participation

Employee participation is a very important element of a TQC effort. Certainly, no TQC effort can be effective without employees contributing to improving products and processes. We have provided suggestions for education in quality methodologies, quality circle activity, and employee suggestion schemes. The result of these activities can be higher morale and productivity and increased customer and employee satisfaction.

Chapter

7

Getting Started and Ongoing Management

Quality is never an accident, it is always the result of an intelligent effort.
                                JOHN RUSKIN

Overview: Why Start a TQC Effort?

There are several reasons to start a companywide total quality control effort. We list here three of them.

1. *To increase profits.* As shown in Chap. 1, a TQC effort will result in less rework, increased productivity, and higher customer satisfaction; the outcome of this can be higher profits. The company's management team will realize this and start the TQC effort.

2. *An enlightened chief executive.* An enlightened chief executive will realize that TQC will lead to a better competitive position and higher profits. The chief executive will then drive the entire company toward his or her quality vision.

3. *A company in a crisis situation.* In this case, the company management realizes that there is a crisis—products don't sell or there is a competitive threat—and adopts a TQC effort.

The most common reason seems to be the third reason. Here are some examples:

- Most Japanese companies started the efforts in quality after 1945, when the economy was ravaged. Japan had to export or perish, but because the word *Japanese* was synonymous with poor quality, they could not export. Through Edward Deming's and Joseph Juran's efforts, which subsequently developed into the body of knowledge known as TQC, Japanese companies improved their quality. Today,

Japanese exports provide such a high balance of trade that it is constantly requested to import more.

- After 1973, when oil prices increased dramatically, Japanese companies adopted a second wave of TQC. Yokogawa Hewlett-Packard, discussed in Chap. 1, was one of those companies.
- In the United States, the Xerox Company launched its TQC effort because of declining market share and won the Malcolm Baldrige National Quality Award. In the late 1980s, the Florida Power & Light Company faced a crisis of deregulation and the aftereffects of the Three Mile Island nuclear scare. It launched its TQC effort and became the first company outside Japan to win the Deming Prize.

Today, reasons 1 and 2, given above, are getting common—partly because management realizes that there is no choice and partly because many large companies require their suppliers to improve quality. Examples are the Deming Prize companies in Japan and all automobile manufacturers in the United States. Our discussion will focus on some of the ways to start and manage a TQC effort.

The Quality Steering Committee

The entire TQC effort must be planned and managed by the company's or organization's management team. There must be a plan and a long-term commitment. The management team must dramatize the importance of quality by making it a strategic issue and by giving quality top consideration. They can demonstrate their commitment by the way they react to quality issues, the kind of people they promote, and the quality goals they set.

Quality should be managed via a quality steering committee or quality council. Membership must be drawn from the senior management team. Such a high level of involvement is necessary if success is to be assured—anything less will lead to failure. Employees of a company or organization learn very quickly how serious you are about quality. They will take the lead not only from your words but also from your actions. Hence, we stress again the need for senior management's participation in leading the quality effort.

The charter and responsibilities of the quality steering committee include the following:

- Establishing a long-term quality vision and goal.
- Establishing the major dimensions or elements of the TQC effort, for example, deciding to establish the Hoshin planning methodology, setting up systems to ensure a customer-obsessed company, and including the initiatives suggested in this text.

- To plan for and provide resources for quality training.
- To provide recognition and rewards to individuals and teams who contribute to the quality effort.
- Reviewing progress toward the quality effort by conducting TQC reviews of the entire organization. This is discussed in Chap. 8.
- Measuring progress toward the organization's vision and goal.

For large companies or organizations, committees or councils will be needed at other levels such as divisions or departments. Even for a small organization, additional subcommittees may be needed to manage organizationwide initiatives such as suggestion scheme activities and quality circle or team activities.

Establishing a quality vision and goal

It is absolutely essential for the company, via the quality steering committee, to establish a quality vision. The vision is a statement of where you want to be in quality. The vision should be specific and have a numerical target which must be achieved in, say, 5 years. We recommend, however, that the quality vision be incorporated in the company's purpose and vision. Refer to the discussion in Chap. 3 in the section "The Long-Range Plan." There, we suggested that the company's purpose and vision include a quality or customer satisfaction statement and that it be supported by a specific goal.

Let us explore some examples of a quality vision

- In the early 1980s, John Young, the Chief Executive Officer of Hewlett-Packard Company, announced the 10 × goal for the entire company. The purpose of this goal was to reduce defects in all hardware products by a factor of 10 within 10 years. This was an aggressive goal, requiring new methodologies. To meet this challenge, Hewlett-Packard launched its companywide TQC effort in 1984 and by the year 1990 the company met this goal.
- In 1986, the Motorola Company, which recently won the Malcolm Baldrige Award, launched its 6 sigma program—a 5-year goal to approach the standard of zero defects. The program focuses on reducing defects in every individual part—whether manufactured or purchased—to a very low, predetermined level by 1992. If all parts have high quality or low fail rates, the assembled products will also be of very high quality. Motorola expects its suppliers to participate in the program.
- More recently, the Xerox Company won the Malcolm Baldrige Award. Since then, it has set a company goal of providing 100 percent customer satisfaction. The satisfaction level is measured via

detailed customer satisfaction surveys conducted throughout the United States. In Xerox's context, 100 percent satisfaction is when a customer gives the company a score of 4 or 5 on a scale of 1 to 5. The Xerox Company is nowhere near this goal, but it has launched a series of quality initiatives at all levels in the company. We mentioned in Chap. 2 that Xerox reputedly ties part of its managers' bonuses to a minimum threshold level in their customer satisfaction survey scores. We also mentioned earlier that we disagree with this method. A better way is to plan for customer satisfaction by managing the processes that provide satisfaction. If you remember, we discussed how to do this in Chap. 2, when we discussed a customer satisfaction model.

The Xerox Company's vision and goal are certainly more outwardly focused than those at Hewlett-Packard or Motorola, which seem focused inward to the company. Employees can understand and relate to the Xerox vision and goal. Certainly, it is much more personal than Motorola's vision of achieving 6 sigma. We recommend an outwardly focused vision and goal that takes your customers' satisfaction into account. This will cause you to focus on the correct systems to improve.

Setting Up a TQC Headquarters Function

For a large organization, a TQC headquarters function is a necessity. For an organization of over, say, 200 people, we propose 2 to 4 people. There are several activities it must manage and facilitate, including the following:

- Define and manage a quality training program for all managers and employees. A suggested list of programs is given in Chap. 6.
- Organization of quality circle, or team, conventions.
- Organization and facilitation of an employee suggestion scheme.
- Management of a customer complaint and feedback system.
- Conduct companywide customer satisfaction survey.
- Publicity, promotion, and reward programs.
- Other activities as defined by the quality steering committee.

Phases of a TQC Effort

Based on our experience, the following are the phases of a TQC effort; each is discussed below:

- Introductory phase
- Acceleration phase

- Cruising phase
- Second acceleration phase
- Second cruising or mature phase

Introduction phase

The introduction phase involves the decision to launch TQC and preparation of the detailed program. The details of the program must be worked out by the quality steering committee and must include preparation of a quality vision and some of the elements discussed in the section on quality vision. The actual launch should be accompanied by some fanfare and publicity. Specifics will include:

- Preparing a quality vision and goal. Preferably this should be woven into the company vision.
- A high degree of management commitment. This means a willingness to devote time and effort and to provide resources.
- The training program, including the PDCA cycle.
- Identification of areas and processes for improvement.
- Identification of specific improvement activities for quality circles and teams.
- Setting up of a TQC headquarters.
- Promotion, publicity, and reward systems.

After training is conducted and quality activities take hold, you will see results such as quality circles, or teams, and a proliferation of improvement projects.

Acceleration phase

As results are seen, activities can be accelerated. The Hoshin planning process should be introduced at this point. It should be used to drive all quality activities as well as to steer the company in the right decision. Specifics here can include:

- Training and use of Hoshin planning and Daily Management, or Business Fundamentals.
- Deployment of Hoshin planning strategies.
- Conduct open and honest Hoshin plan reviews.
- Select cross-functional projects in the Hoshin plan. These projects cross functional and department boundaries.
- Introduction of an employee suggestion scheme.
- Identify specific processes for improvement as suggested in Chap. 5.

- Continue with promotion, publicity, and reward systems.
- Begin to pursue customer issues such as customer complaints and customer satisfaction surveys.
- Continue with senior management support and commitment.

Cruising phase

When the first and second phases are well managed, you will get to the cruising stage. If not, you may have collapsed and given up. You will reach this stage after several years. Once here you may be exhausted. After some time here, there is danger of falling back or reaching a feeling of burn out. This is a common phenomenon that we have observed. In this stage, you consolidate and fine-tune all the activities and programs launched. Specifics include coverage of the following:

- Ensure employees and managers are customer obsessed.
- Ensure good planning process.
- Ensure good improvement process (use and understanding of PDCA cycle).
- Ensure that key processes are managed.
- Measure progress toward the quality vision and goal and take corrective action as needed.
- As fine-tuning continues, you are ready to conduct TQC reviews—discussed in the next chapter. These reviews will emphasize and restate the management commitment that already exists. It will also raise awareness of challenges and opportunities.

Finally, when this phase is managed well, you can start a second acceleration phase.

Second acceleration phase

In the cruising stage, despite the best efforts, you will get complacent. The introduction of TQC reviews will help reaccelerate your progress; reviews will also help to identify and eliminate weak spots and departments. But in addition, a real shot in the arm can come from applying for the Overseas Deming Prize, the Malcolm Baldrige Award, or other awards available in your country.

In this phase, you should be doing less quality improvements and more quality creation. Here you are in a problem prevention by prediction phase. (Refer to the discussion in Chap. 4 of the problem-solving hierarchy.) *In addition, you are in the second zone of TQC—total quality creation.* All your products and services should have attractive quality features. Refer to the discussion of attractive quality in Chaps. 1 and 2.

Second cruising phase

The danger of collapse and complacency is much less in the second cruising phase. Continue with TQC reviews. Get external world-class consultants to conduct some of the reviews to inject more vitality into your TQC efforts. You should have a first-class TQC headquarters by now—part of its role is to keep the momentum going and to inject fresh ideas and vigor into the organization. *It is especially important that the organization does not become complacent with success, and management must be wary of this. Complacency is the very enemy of quality.*

TQC and Time Management

One of the constant complaints of managers is that there is no time to do TQC. We feel TQC is a better way to manage, and we would like to show the relationship between the elements of TQC and a manager's basic activities.

The basic activities of a manager

For the answer to this question, let us look at a time management model from Itoh.* This is shown in Fig. 7.1. Let us discuss this model

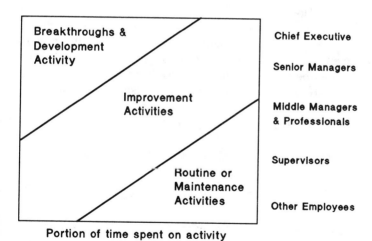

Figure 7.1   The Itoh model.

---

*We are not able to trace the source of the Itoh model; we have seen it many times without reference to its source. We have seen numerous variations of this, including the one used by the Kaizan Institute of America. The Itoh model predates the model used by the Kaizan Institute.

for a while. You read it from left to right. The specific activities are shown in the box, and as you glance to the right, you see the layer of management or employees that is responsible for that activity. What are the main activities in an organization and who does them?

Breakthrough or development activity. This is the primary activity for a successful organization. This includes new development and introduction of new products and services; taking the steps to steer the organization in the correct direction to the right destination; investments to grow, develop, and strengthen the organization; and so on.

The responsibility for this activity rests with the following:

- Clearly this activity is the work of senior managers. In fact, as shown by the segmentation of the activity, the bulk of this is done by senior managers.
- This is probably the only portion of the TQC in which the chief executive of the organization plays an active role and rightly so. If he or she does many other things, it may indicate that time is not optimally used.
- Many professionals, such as engineers, would also be involved in this activity but mostly in creation, preparation, and design of new products and services.

Improvement activity. Continuous improvement is another important activity. An organization cannot rely on a continuous stream of blockbuster products and services. Many of the current processes, products, and services will require improvement—be it an automobile, a computer, a turbine engine, an instrument, or business-class service on an airline. These cannot be simply replaced with a brand new offering—so you improve them. Key internal processes will also require improvement: processes such as new product development, manufacturing, inventory management, supplier management, marketing, sales, delivery, or distribution. Where possible, benchmark them against the competition and develop aggressive goals for improvement. The following are responsible for this activity:

- Senior management's role is to identify, set priorities, and review progress.
- The bulk of the work will be done by middle managers and professionals, such as engineers and accountants. They will form project teams, quality teams, and cross-functional task forces to facilitate this important activity.
- Supervisors and all the employees reporting to them will be involved. Some of the activities will be in improving key internal processes as identified by senior management, but other activities will

be at the micro level—lower-level quality circles working on processes in their environment, for example, increasing the yield at a specific assembly point or reducing paperwork.

Routine or maintenance activity. This refers to management of routine day to day processes. These processes must be maintained at their current level of performance. This would include things like the entire assembly line in an automobile, computer, or turbine engine factory; assembly in a diaper or cereal factory; or all the routine lending and borrowing activity in a bank.

These activities represent the fundamental processes of an organization—processes which are running routinely and require little management intervention, so long as they are in control. Responsibility for this activity is as follows:

- This activity is done primarily by the workers and their supervisors.
- Supervisors who manage the employees focus mostly on this activity and a little on the improvement activity as shown in Fig. 7.1.
- There may be some involvement by middle managers and professionals—such as emergency intervention and fine-tuning of the maintenance activity, but involvement should be minimal. If it is high, there is a possibility that the activity is not routine—it may indicate fire fighting to fix unstable processes.

Fitting the TQC elements in the Itoh model

In the last five chapters we have discussed the five elements of TQC, specifically:

- Customer obsession
- Planning process
- Improvement cycle
- Process management
- Employee participation

Within these elements we have discussed numerous quality tools and methodologies. Those tools can and will help manage the three critical activities shown in the Itoh model. In Fig. 7.2 this model has been overlain with the five elements of TQC. Look at the figure. You can see that the various elements of TQC can help to manage the three activities. For example:

1. *Breakthroughs or development activity.* Use Hoshin and long-range planning to help in the planning of new products and services and in steering the organization in the right direction. In addition,

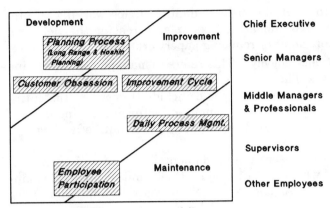

Portion of time spent on activity.
Figure 7.2 The TQC elements overlaid on the Itoh model.

the techniques mentioned in Chap. 2, "Customer Obsession," can help in identifying areas for possible breakthroughs in products, services, or processes. Some of the techniques are competitive benchmarking and capturing the customer's voice and needs.

2. *Improvement activity.* Use Hoshin and long-range planning to help plan for what needs to be improved. Information on selecting items to be improved will come from the various techniques mentioned in Chap. 2, "Customer Obsession," such as a customer complaint and feedback system and customer surveys. Processes to improve have been discussed in Chap. 5, "Daily Process Management." Use the PDCA improvement cycle to manage the improvement. The various employee participation activities, such as quality circles and suggestion schemes, fit well in this segment.

3. *Routine or maintenance activity.* The key processes that must be routinely managed can be listed and monitored via the Daily Management, or Business Fundamentals, plan. These were discussed in Chap. 3 and include key processes in design, manufacturing, sales, service, and postmortems. Each of these processes must be properly documented, everyone must be well trained, and the process must be well managed.

As you can see, TQC should not be an additional activity for managers but instead it should be used for managing the three key activities in an organization that are identified in the Itoh model. The three activities listed are crucial for management and employees, and the TQC elements can help to manage these activities.

Integrating TQC effort into management activity, or making TQC a way of life. The initial TQC effort will start with the TQC steering committee's effort, but as quickly as possible, the TQC effort must be made a way of life. To do this well, we recommend that the elements of TQC be part of the planning process (see Fig. 7.3).

In Fig. 7.3, we identify the elements of TQC and show when to plan for them. For example, planning starts with the long-term plan, customer inputs, and competitive data. Here the annual Hoshin plan and the Daily Management, or Business Fundamentals, plans are prepared. At this point, employee participation and daily process management issues are reviewed. During the planning stage, items for improvement are identified; these are managed during the rest of the year. Finally, all these activities are reviewed for progress during the coming year. By integrating the tools and techniques of TQC into the planning process, we can make TQC a way of life at any organization.

Some Questions and Answers on Getting Started and Some General Questions on TQC

Q1. I run a small independent operation of about 100 people; do you expect me to have a TQC headquarters and a quality manager?

A1. For a small organization of less than 100 people, it will not be possible to justify having a quality manager or a TQC headquarters. We recommend you drive TQC in your organization via a TQC steering committee. If you have about 100 people or more, you should be able to justify having a quality manager to facilitate TQC activity. And once you are past 200 people, you can consider adding more resources.

Q2. We started our TQC efforts many years ago, but the effort continues to move slowly and many employees are unconvinced that it will help them. What's wrong? Also, why does a TQC effort fail in some organizations?

A2. The most common reasons why a TQC effort fails are lack of management commitment, involvement, and impatience. If the chief executive and his or her staff believe in it and drive the effort, there will be success. This is why most Japanese companies succeed and win the Deming Prize. This is also why Florida Power & Light and Xerox have been successful in the United States. So if progress is slow, go back and examine your actions, behavior, and commitment.

What is commitment? It includes changing the policies of the company by adopting the five elements of TQC. This means making decisions that are right for the long term, although in the short term, they will be tough and difficult.

Q3. Why bother with starting a TQC program? We can be more competitive by cutting costs through laying off excess and unproductive employees.

A3. Cost cutting by massive employees layoffs can be effective, but

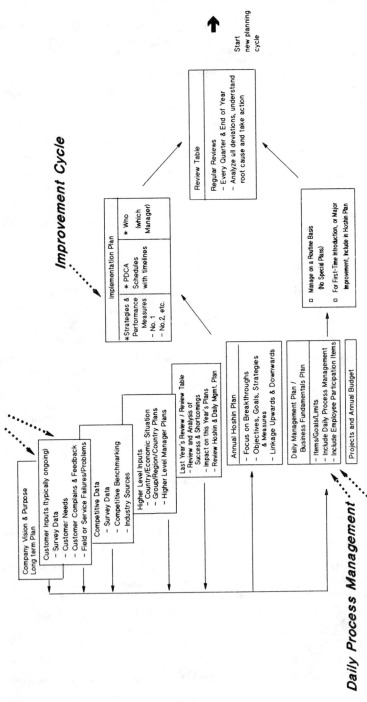

Figure 7.3 Integrating the elements of TQC into the planning process. This figure illustrates the planning process we discussed in Chap. 3. Note that the other elements of TQC, such as customer obsession, improvement cycle, daily process management, and employee participation, are integrated into planning.

it only gives short-term results because it does not improve inefficient processes. A TQC effort on the other hand, gives long-term results by providing efficient and productive processes, less waste, better products and services, and a high degree of customer focus. In addition, if you use the Hoshin planning process correctly, there is ongoing cost control. This means a lower likelihood that you will allow costs to get out of control.

Q4. Our quality is already very good; why do we need to start a TQC effort?

A4. It is possible that your quality is very good—you may even be considered a world leader in the products or services you offer. But quality is a moving target. Today's leader can be tomorrow's follower. Just witness what happened in the automobile, consumer electronics, and camera industries. The list is endless. A leader is in a dangerous position because competitors will want the leadership position. All the methodologies offered in this text help to keep you on your toes and moving ahead.

Summary: Getting Started and Ongoing Management

We have discussed how you should get started in a TQC effort and how to ensure effective ongoing management. For any TQC effort to succeed, first you need a quality vision and a TQC steering committee to define and guide the effort. Progress must be monitored, and if there are roadblocks, they must be removed. In the next chapter, we will discuss how to review a TQC effort and measure progress. A TQC headquarters will be essential in facilitating the process and in injecting fresh ideas and vigor.

TQC methodology is not an alternative method of management nor is it additional work or a separate program. TQC methodology provides tools that allow you and your organization to be more effective and to better manage current activity.

The Itoh time management model gives us the three main activities for management and its employees. These are:

- Development and breakthrough activity
- Improvement activity
- Routine or maintenance activity

The TQC elements we have discussed can be used to enhance performance in these activities and lead to a more successful and profitable organization. These TQC elements must be planned for and work best if integrated into the annual planning process.

Finally, for the TQC effort to start and succeed, management commitment and involvement is essential. With success may come complacency and that must be avoided because TQC is a never ending journey for continuous improvement and betterment of life.

Chapter

8

Conducting TQC Audits, or Reviews

It is obvious that such reviews, conducted by upper level managers, can have a powerful impact throughout the company. The subject matter is so fundamental in nature that the reviews can reach into every major function.
JOSEPH M. JURAN

I learned more about my division in 2 days of the TQC review than in my total 9 months tenure as a General Manager.
A NEW HEWLETT-PACKARD GENERAL MANAGER

Overview

Why do a TQC audit, or review? To answer this question, let us look at an analogy in daily life. We often go for medical checkups. The purpose of and requirements for a medical checkup are well understood. The doctor gives a diagnosis by measuring performance against what is possible. When we hear the results, we decide what to do with the recommendations. The result can be a better and healthier life.

The purpose of TQC audits, or reviews, is similar to medical checkups. During the review a diagnosis is provided, and if the recommendations are followed, they can lead to a more efficient and productive operation. In TQC, however, the requirements or standards are not well known—but in this text, we have defined the minimum requirements. We will provide detailed checklists and guidelines for conducting a TQC review.*

---

*The TQC review procedure discussed here is a modification of the Quality Maturity System, developed by the author at the Hewlett-Packard Company, Palo Alto, California.

Deming Prize and Malcolm Baldrige Award Criteria

Before we provide a proposed TQC review process, let us take a quick look at the criteria used by the Deming Prize and Malcolm Baldrige Award Committees.

Deming Prize criteria

The Deming Prize Committee, based in Japan, looks at 10 categories when assessing an applicant for the Deming Prize. We list the categories below with a detailed breakdown for each category. The list comes from the English translation of the checklist for the Deming Prize application.

1.0. *Policy.*
  1.1. Policies pursued for management, quality, and quality control
  1.2. Method of establishing policies
  1.3. Justifiability and consistency of policies
  1.4. Utilization of statistical methods
  1.5. Transmission and diffusion of policies
  1.6. Review of policies and the results achieved
  1.7. Relationship between policies and long- and short-term planning

2.0. *Organization and its management.*
  2.1. Explicitness of the scopes of authority and responsibility
  2.2. Appropriateness of delegations of authority
  2.3. Interdivisional cooperation
  2.4. Committees and their activities
  2.5. Utilization of staff
  2.6. Utilization of QC circle activities
  2.7. Quality control diagnosis

3.0. *Education and dissemination.*
  3.1. Education programs and results
  3.2. Quality and control consciousness, degrees of understanding of quality control
  3.3. Teaching of statistical concepts and methods and the extent of their dissemination
  3.4. Grasp of the effectiveness of quality control

3.5. Education of related company (particularly those in the same group, subcontractors, consignees, and distributors)

3.6. QC circle activities

3.7. System of suggesting ways of improvements and its actual conditions

4.0. *Collection, dissemination, and use of information on quality.*

4.1. Collection of external information

4.2. Transmission of information between divisions

4.3. Speed of information transmission (use of computers)

4.4. Data processing, statistical analysis of information, and utilization of the results

5.0. *Analysis.*

5.1. Selection of key problems and themes

5.2. Propriety of the analytical approach

5.3. Utilization of statistical methods

5.4. Linkage with proper technology

5.5. Quality analysis, process analysis

5.6. Utilization of analytical results

5.7. Assertiveness of improvement suggestions

6.0. *Standardization.*

6.1. Systematization of standards

6.2. Method of establishing, revising, and abolishing standards

6.3. Outcome of the establishment, revision, or abolition of standards

6.4. Contents of the standards

6.5. Utilization of statistical methods

6.6. Accumulation of technology

6.7. Utilization of standards

7.0. *Control.*

7.1. Systems for the control of quality and such related matters as cost and quantity

7.2. Control items and control points

7.3. Utilization of such statistical control methods as control charts and other statistical concepts

7.4. Contribution to performance of QC circle activities

7.5. Actual conditions of control activities

7.6. State of matters under control

8.0. *Quality assurance.*

    8.1. Procedure for the development of new products and services (analysis and upgrading of quality, checking of design, reliability, and other properties)

    8.2. Safety and immunity from product liability

    8.3. Process design, process analysis, and process control and improvement

    8.4. Process capability

    8.5. Instrumentation, gauging, testing, and inspecting.

    8.6. Equipment maintenance and control of subcontracting, purchasing, and services

    8.7. Quality assurance system and its audit

    8.8. Utilization of statistical methods

    8.9. Evaluation and audit of quality

    8.10. Actual state of quality assurance

9.0. *Results.*

    9.1. Measurement of results

    9.2. Substantive results in quality, services, delivery time, cost, profits, safety, environment, etc.

    9.3. Intangible results

    9.4. Measures for overcoming defects

10.0. *Planning for the future.*

    10.1. Grasp of the present state of affairs and the concreteness of the plan

    10.2. Measures for overcoming defects

    10.3. Plans for future advances

    10.4. Linkage with the long-term plans

Malcolm Baldrige National Quality Award

Next, let's take a look at the criteria for the Malcolm Baldrige National Quality Award. The list also shows the maximum points awarded in each category and is taken from the brochure provided by the examination committee.

1990 examination categories/items		Maximum points
1.0. Leadership		100
1.1. Senior executive leadership	30	
1.2. Quality values	20	
1.3. Management for quality	30	
1.4. Public responsibility	20	
2.0. Information and analysis		60
2.1. Scope and management of quality data and information	35	
2.2. Analysis of quality data and information	25	
3.0. Strategic quality planning		90
3.1. Strategic quality planning process	40	
3.2. Quality leadership indicators in planning	25	
3.3. Quality priorities	25	
4.0. Human resource utilization		150
4.1. Human resource management	30	
4.2. Employee involvement	40	
4.3. Quality education and training	40	
4.4. Employee recognition and performance measurement	20	
4.5. Employee well-being and morale	20	
5.0. Quality assurance of products and services		150
5.1. Design and introduction of quality products and services	30	
5.2. Process and quality control	25	
5.3. Continuous improvement of processes, products, and services	25	
5.4. Quality assessment	15	
5.5. Documentation	10	
5.6. Quality assurance, quality assessment, and quality improvement of support services and business processes	25	
5.7. Quality assurance, quality assessment and quality improvement of suppliers	20	

(Continued)

1990 examination categories/items	Maximum points
6.0. Quality results	150
6.1. Quality of products and services — 50	
6.2. Comparison of quality results — 35	
6.3. Business process, operational and support service quality improvement — 35	
6.4. Supplier quality improvement — 30	
7.0. Customer satisfaction	300
7.1. Knowledge of customer requirements and expectations — 50	
7.2. Customer relationship management — 30	
7.3. Customer service standards — 20	
7.4. Commitment to customers — 20	
7.5. Complaint resolution for quality improvement — 30	
7.6. Customer satisfaction determination — 50	
7.7. Customer satisfaction results — 50	
7.8. Customer satisfaction comparison — 50	
Total points	1000

Comparison between the Deming Prize and the Malcolm Baldrige Award

Both the Deming Prize and Malcolm Baldrige Award are good because they provide a detailed checklist and set expectations for a customer- and quality-oriented organization. Any company that pays attention to the checklists is bound to improve.

The Malcolm Baldrige Award provides a very detailed checklist and more details can be obtained from the award committee. It also gives a very detailed scoring system that can be useful in self-assessment. Its flaw is that it is results-oriented and does not focus sufficiently on the process, or way of doing things.

We believe that not focusing sufficiently on the process is a major flaw. It reflects on the western management concept of focusing on results—often short-term results. Or as Edward Deming often says in his lectures: Western managers get results in any way—*"mind not required."* As a result, it is possible to win the Malcolm Baldrige Award by bulldozing toward specific goals. Later, however, an award-winning company may not hold onto the gains; instead it could lose many of the gains.

In our discussion on the essence of TQC in the next chapter, we emphasize that the process and PDCA are a crucial part of TQC because they ensure a robust organization, an organization where quality improvement gains remain in place and results are predictable even as managers come and go.

The Deming Prize checklist focuses on both the process and results. There are at least three categories that look at the process: analysis, standardization, and control. The Deming criteria, however, does not give a detailed scoring method. As far as we are aware, none exists. The examining board has experienced and knowledgeable experts who rely on their experience and discussions to grade an applicant. This provides extreme flexibility and just about any type of organization can be graded—manufacturing or service. Nevertheless, a detailed scoring system would be very useful. Because it does not exist, an applicant for the Deming Prize usually works with a registered consultant from the Japanese Union of Scientists and Engineers (JUSE) in preparing for the application.

A TQC Review Procedure

The objective of a TQC review is to increase business success by building a stronger and more competitive organization through the use of TQC methodology. This is done by:

- Reviewing and diagnosing the process and skills used in managing the organization
- Identifying strengths, weaknesses, and opportunities for improvement

The TQC review process

The TQC review process consists of:

- An agenda for the review
- A detailed checklist for the review, with a scoring system
- A trained review team to conduct the review
- The TQC review report with recommendations for improvement

We will discuss each of these items in detail and provide guidelines and recommendations.

An agenda for the review

We will provide three agendas for conducting a review. These are:

1. An agenda for reviewing a design and manufacturing entity (a product division or operation)
2. An agenda for reviewing a sales entity (a sales and marketing operation)
3. An agenda for reviewing a single department in a design, manufacturing, or sales entity

The suggested timings and areas to be reviewed have been developed after conducting numerous reviews of manufacturing and sales entities. If you decide to make changes, we suggest you do so only after trying this process.

Agenda no. 1. The following is a sample agenda for a manufacturing entity:

Item reviewed	Who or Function	Length (hours)
Opening remarks by review team (TQC review objectives, role of review team)	Team leader	0.25
Overview (entity mission, markets and customers, products and services, business environment, and current performance)	General manager	0.25
Customer obsession	Quality, marketing, R&D	3.00
Planning process	General manager's staff	3.00
Review of first department		
Department plan review	Manufacturing	0.50
Improvement cycle	Manufacturing	1.50
Process management	Manufacturing	2.00
Review of second department		
Department plan review	R&D	0.50
Improvement cycle	R&D	1.50
Daily process management	R&D	2.00
Total participation	General manager, quality	1.00
Break (review team prepares comments)		1.00

(Continued)

Item reviewed	Who or Function	Length (hours)
Review team gives verbal report (observations, recommendations for improvement, explanation of scoring procedures)	General manager's staff	1.00
Total time (approximately)		17.00

Agenda no. 2. The following is a sample agenda for a sales and marketing entity:

Item reviewed	Who or Function	Length (hours)
Opening remarks by review team (TQC review objectives, role of review team)	Team leader	0.25
Overview (entity mission, markets and customers, products and services, business environment, and current performance)	General manager	0.25
Customer obsession	Sales, marketing, customer support, quality	3.00
Planning process	General manager's staff	3.00
Review of first department		
Department plan review	Sales and marketing	0.50
Improvement cycle	Sales and marketing	1.50
Process management	Sales and marketing	2.00
Review of second department		
Department plan review	Customer support or service	0.50
Improvement cycle	Customer support or service	1.50
Daily process management	Customer support or service	2.00
Total participation	General manager, quality	1.00
Break (review team prepares comments)		1.00
Review team gives verbal report (observations, recommendations for improvement, explanation of scoring procedures)	General manager's staff	1.00
Total time (approximately)		17.00

An important note for agendas 1 and 2. When an entity is reviewed again, progress against the deficiencies noted in the previous review will be measured. Hence, it will be appropriate to start the review with a discussion on improvements since the last review. Formal reviews can be conducted every 2 to 3 years. Informal entity reviews, or self-audits, should be conducted annually by your quality manager. In-depth reviews of the entity's major departments should also be conducted annually by the quality manager, as per agenda no. 3.

Agenda no. 3. The following is a sample agenda for reviewing a single department, or large function (such as R&D, manufacturing, or sales), at any type of entity.

Item reviewed	Who or Function	Length (hours)
Opening remarks by review team	Team leader	0.25
Customer obsession	Department manager's staff	2.00
Planning process	Department manager's staff	2.00
Improvement cycle	Section managers	1.50
Daily process management	Section managers	1.50
Total participation	Department manager's staff	0.75
Review team gives verbal report	Department manager's staff	1.00
Total time (approximately)		9.00

Detailed checklist for the review

The checklist will require two dimensions: a list of pertinent questions and a method of scoring the questions. For scoring, we recommend a system based on the concepts used in the Malcolm Baldrige Award scoring system. Let's start with the scoring system and then proceed to the checklist. Note that in developing the scoring system and checklist, we have tried to keep everything short, simple, and usable in a wide area.

Scoring system

The scoring matrix illustrated in Fig. 8.1 is designed to highlight an entity's strengths and weaknesses and to facilitate feedback. It is intended to help monitor progress, which will come from conducting regular TQC reviews.

Not using data or quality management tools. No evidence of effort. No results or evidence of quality.	Using some data and quality management tools. Evidence of effort in a few areas. Numerous opportunities exist. Little success, lots of opportunity.	Knowledgable use of data and quality management tools. Evidence of effort in several areas. Further deployment possible. Some success, but more needed.	Good use of data and quality management tools. Some innovative approaches. Evidence of effort and deployment in most areas. Successful in most areas with some areas requiring more attention.	Excellent use of data and quality management tools. Many innovative approaches. Evidence of effort and deployment in all areas. Good to excellent results in all areas.
Score: 0	1	2	3	4

Figure 8.1 The scoring matrix.

Guidelines for using the scoring system. When you look at each question in the checklist, select the cluster of comments that is most appropriate. Then choose the score associated with the cluster of comments—this is the score for the question.

Choosing a score value of half a point. Sometimes, you may feel that the score falls in between two clusters, for example, between 0 and 1. In this case, choose a score value of 1/2, or 0.5.

Choosing a score value of 0 or 4. A score value of 0 will be very obvious to you when you come across it. A value of 4 represents world class and should be used only when it is clearly appropriate. If in doubt, instead of a value of 4, give a lower value. We expect that only a world-class or extremely experienced reviewer—one who has reviewed or visited several world-class companies will be able to give a score value of 4.

Score value for multiple questions. Several questions in the checklist are actually multiple questions. Faced with a choice of having a long list of questions or a short list, we choose the latter. This has caused many questions to be grouped into clusters. This is what we suggest you do: *If all items in a multiple question are done extremely well and are done in all areas, give a score of 4.* If only one item (out of, say, three items) is done well, give a score of one-third of that, or 1.3. On the other hand, if one department does all items in a multiple question well, but two other departments do not do it at all, give a score of, say, 1 instead of 3. As an example, look at question 5 in the customer-obsession section. If good competitive data has been obtained but no analysis or follow-up, give a score of 1.3.

Summarizing the scores. The summary of the score for each element can be displayed on a radar chart, with five axes. Each axis represents one of the elements of TQC. In our proposed system, we retain the scale of 0 to 4 for each TQC element.

The summary radar chart is shown in Fig. 8.2. This is a convenient mechanism for displaying long-term targets (for example, 3.0 for all elements) and displaying actual scores for successive TQC reviews. A similar system exists at the Hewlett-Packard Company.

An alternate system to display results and include comments on the overall quality maturity is shown in Fig. 8.3 (p. 266). This is based on a similar method that is used by the Malcolm Baldrige Award Committee.

Which should you use? Both have their merits, and we suggest using both. The radar chart allows you to display specific areas of weakness and measure progress over the years. The scoring comments provide a description of the current quality status.

Scoring range. We have used a scoring range of 0 to 4. If you prefer you can use a range of 1 to 5 or even 1 to 10 (if you dislike half points). All you need to do is to change the numbers in Fig. 8.2 and make minor changes to Figs. 8.2 and 8.3 and to the score sheets. If you need help, contact the author.

Review checklist

We will now give a review checklist. The format is as follows for each TQC element:

- Checklist of questions. These have been kept to five or seven questions per element. However, we have structured many of these as multiple questions. This helps to ensure simplicity and consistency.
- Where appropriate, a short discussion is given after each question that sets expectations.
- A separate page allows scoring of each element. This includes the scoring matrix, discussed earlier.

Conducting TQC Audits, or Reviews 253

Once scores are determined for each category and calculated, they will be displayed as a radar chart like the one shown below. The radar chart will give managers a snapshot of how their entity scored in each area.

1. Planning Process   _____   (maximum 4)

2. Customer Obsession   _____   (maximum 4)

3. Improvement Cycle   _____   (maximum 4)

4. Process Management   _____   (maximum 4)

5. Total Participation   _____   (maximum 4)
   (Includes Employee Participation)

   Total Score: _____   (maximum 20)

   OVERALL SCORE:
   To compute: Total Score   (maximum 4)
                ──────────   _____
                     5

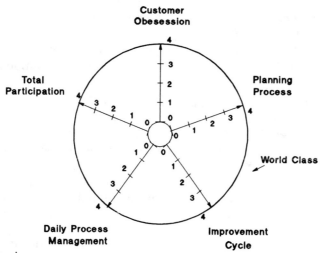

Figure 8.2 Scoring summary.

Customer-obsession checklist. In this section, we review the entity's obsession with customers and management of customer issues, as follows:

1. Customers are identified and stratified.
    Discussion: Here we look at the entity's external and internal customers and determine if they are defined and stratified by needs, products, and services. This item is so obvious for any department, yet many will do poorly here.
2. Customer needs and satisfaction levels determined and follow-up corrective action taken on products and services.
    Discussion: Here we look at the process of determining customer needs and satisfaction levels, an analysis of the available data, and how this analysis influences current products and services and future plans. Items to look for include surveys, customer focus groups, and customer feedback via surveys and/or response cards after purchase.
3. Customer interface and follow-up activities are well established.
    Discussion: Here we determine methods of visiting and meeting with customers to get their inputs, concerns, and suggestions; what is done with the data; and how analysis of this data influences current products and services and future plans.
4. Product and service quality data collection processes are well established. Data is used to drive improvement.
    Discussion: Here we look at the process of collecting failure data on products and services from customers after the sale and at how analysis of this data influences products and services and future plans. Items to look for include product failure information and the customer complaints and feedback system.
5. Competitive data are available and analyzed; the analysis is used to influence both current and new products and services. This will include a sales won or lost analysis and follow-up, corrective, action.
    Discussion: Here we look at competitive benchmarking and other means to get competitive data. For a sales and marketing department, we would look at a sales won or lost analysis, including an understanding of why sales are won or lost and follow-up corrective action.

Note: For all the above items, it is imperative that we find out what is done with the collected data.

Conducting TQC Audits, or Reviews   255

Customer-obsession scoring. This page is set aside for scoring the items reviewed. Use the scoring matrix below and questions provided for determining the scores.

	Score
1. Customers are identified and stratified.	_____
2. Customer needs and satisfaction levels are determined and follow-up action is taken on products and services.	_____
3. Customer interface and follow-up activities are well established.	_____
4. Product and service quality data collection processes are well established. Data is used to drive improvement.	_____
5. Competitive data is available and is analyzed and used to influence current and new products and services. include sales won or lost analysis where appropriate.	_____
Total score: (Maximum 20 points)	_____
Rescale: Total score /5 (Maximum 4 points)	_____

Not using data or quality management tools. No evidence of effort. No results or evidence of quality.	Using some data and quality management tools. Evidence of effort in a few areas. Numerous opportunities exist. Little success, lots of opportunity.	Knowledgable use of data and quality management tools. Evidence of effort in several areas. Further deployment possible. Some success, but more needed.	Good use of data and quality management tools. Some innovative approaches. Evidence of effort and deployment in most areas. Successful in most areas with some areas requiring more attention.	Excellent use of data and quality management tools. Many innovative approaches. Evidence of effort and deployment in all areas. Good to excellent results in all areas.
Score: 0	1	2	3	4

Scoring matrix.

Planning process checklist. In this section, we review the planning process at the entity being reviewed.

1. There is a long-term plan to shape the future of the entity.

    Discussion: This plan should cover items such as the entity's vision, customer needs, products and services, core competencies, competitive situation, partners, financial analysis, and potential threats. Items from this plan drive the annual Hoshin plan.

2. There is an annual planning process with the following features:
    - Generation of key issues and, from them, a few breakthrough objectives.
    - Influenced by previous year's issues and lessons learned.
    - Quality, cost, education, and delivery issues have been addressed.
    - Customer issues are addressed and linked to the customer obsession checklist.

    Discussion: Review the planning process and check for the features mentioned above. If quality, cost, delivery, education, or customer issues are not addressed, an analytical explanation is available.

3. The annual Hoshin plans are robust and include the following:
    - A few breakthrough objectives, with robust supporting strategies.
    - Goals and performance measures are measurable.
    - Higher level plans are properly deployed down the organization.

    Discussion: When plans are reviewed, focus on one or two objectives, considered the most crucial, then check for the above featuresthroughout the plans. Be especially wary if management has too many objectives.

4. Cross-functional issues have been addressed in the annual Hoshin plan.

    Discussion: Check on how cross-functional issues have been addressed, monitored, and managed.

5. Implementation plans exist for the annual Hoshin plans; plans are reviewed and ownership, time, and resources are defined.

    Discussion: Here we look at the lower levels in the organization and review their implementation plans.

6. A robust Daily Management or Business Fundamentals plan exists and includes regular tracking of performance.

   Discussion: Every department must have such a plan. Remember to ensure that the department reviewed distinguishes breakthrough (Hoshin) from daily management issues.

7. All plans are reviewed regularly for progress and corrective action taken, including changes to the plan, when necessary.

   Discussion: Check the review process, progress to plans, lessons learned, and follow-up corrective action.

Planning process scoring. This page is set aside for scoring the items reviewed. Use the scoring matrix below and questions provided for determining the scores.

	Score
1. There is a long-term plan with the features mentioned on the opposite page.	
2. A planning process, with the important features mentioned on the opposite page, is in place.	
3. The annual plans are robust and include the items mentioned on the opposite page.	
4. Cross-functional issues have been addressed in the annual Hoshin plan and are properly managed.	
5. Implementation plans to support the annual plans exist; plans are reviewed and ownership, time, and resources are defined.	
6. A robust Daily Management or Business Fundamentals plan exists and includes regular tracking of performance.	
7. All plans are reviewed regularly for progress and corrective action taken, including changes to the plan, when necessary.	
Total Score: (Maximum 28 points)	
Rescale: Total score/7 (Maximum 4 points)	

Not using data or quality management tools. No evidence of effort. No results or evidence of quality.	Using some data and quality management tools. Evidence of effort in a few areas. Numerous opportunities exist. Little success, lots of opportunity.	Knowledgable use of data and quality management tools. Evidence of effort in several areas. Further deployment possible. Some success, but more needed.	Good use of data and quality management tools. Some innovative approaches. Evidence of effort and deployment in most areas. Successful in most areas with some areas requiring more attention.	Excellent use of data and quality management tools. Many innovative approaches. Evidence of effort and deployment in all areas. Good to excellent results in all areas.
Score: 0	1	2	3	4

Scoring matrix.

Improvement cycle checklist. In this section, we review several improvement projects—at least two per department reviewed.

1. Problem definition and linkage to the annual plan or crucial issue.

   Discussion: Improvement projects are linked to the annual plan or crucial issues. Problems, definitions, processes, improvement goals, and project schedules are defined and available.

2. Data collection, analysis, and understanding of root cause.

   Discussion: Good collection of data and appropriate use of quality tools such as Pareto and cause and effect diagrams and graphs. Most likely causes have been identified and verified with data.

3. Alternative solutions reviewed and results achieved. Deviations from goal are understood.

   Discussion: Alternative solutions are reviewed, evaluated, and implemented. Results have been achieved and measured; deviations from goals are well explained and acted upon.

4. Review standardization and future plans. New process is standardized and employees are trained.

   Discussion: New or modified processes are documented and employees are trained; future plans exist. Look for a method to update standards across the organization, for example, a method similar to the SUR format, discussed earlier.

5. Adherence to the PDCA cycle. Systematic approach used with no recurrence of problem.

   Discussion: During review of the improvement cycle, check for adherence to the steps of the PDCA cycle and documentation of the entire project. *Most important, look for lessons learned and reasonable assurance that this problem will not recur at the entire entity—you must probe extensively for this.*

Improvement cycle scoring. This page is set aside for scoring the items reviewed. Use the scoring matrix below and the questions provided for determining the scores.

	Score
1. Problem definition and linkage to annual plan or crucial issues.	___
2. Data collection, analysis, and understanding of root cause.	___
3. Alternative solutions reviewed and results achieved. Deviations from goal are understood.	___
4. Review standardization and future plans. New process is standardized and employees are trained.	___
5. Adherence to the PDCA cycle. Systematic approach used with no recurrence of the problem.	___
Total Score: (Maximum 20 points)	___
Rescale: Total score/5 (Maximum 4 points)	___

Not using data or quality management tools. No evidence of effort. No results or evidence of quality.	Using some data and quality management tools. Evidence of effort in a few areas. Numerous opportunities exist. Little success, lots of opportunity.	Knowledgable use of data and quality management tools. Evidence of effort in several areas. Further deployment possible. Some success, but more needed.	Good use of data and quality management tools. Some innovative approaches. Evidence of effort and deployment in most areas. Successful in most areas with some areas requiring more attention.	Excellent use of data and quality management tools. Many innovative approaches. Evidence of effort and deployment in all areas. Good to excellent results in all areas.
Score: 0	1	2	3	4

Scoring matrix.

Daily process management and standards checklist. In this section, we review the concept of standards and at least two key processes in each department. A list of processes was shown in Fig. 5.5, and a shorter list is given here. When other processes are reviewed or if an area not listed in our list is reviewed, determine if the process reviewed is a key process—is it fundamental to success or can it be eliminated? In most cases, the process will be listed in the Daily Management plan of the department being reviewed. If not, question why you are reviewing it.

1. Is the concept of standards well understood?

    Discussion: Look for the current list of standards and adequate documentation. Check for the following:

    - An entity quality manual that documents the overall quality assurance system (discussed in Chap. 5). Check for details on standards and owners and data on implementation. The items in the manual should include the items shown in Fig. 5.19. This would be applicable in design, manufacturing, and sales organizations. Other types of organizations should have something very similar—that is, a quality manual that documents how they manage the quality of their processes, products, and services.
    - Documentation on other standards used at the entity, for example, planning, design, accounting, sales, and marketing.
    - For the items listed above, how are the standards established, revised, and abolished?

Note: This question and those that follow can be modified to review the requirements of ISO 9000, as listed by the International Standards Organization. This is fast becoming a requirement in the European Economic Committee (ECC).

2. The key process is identified, documented, and understood.

    Discussion: There is a flowchart, steps are documented, there are performance measures and targets, and employees have been trained. Ongoing training is provided to compensate for employee turnover and process changes.

3. Key process is monitored and performance against targets (and action limits) is checked routinely.

    Discussion: The process is monitored routinely. For a process such as a quality assurance system and quality checkpoints (Table 5.5) or sales process management (Fig. 5.24), performance measures have been established, with targets. For a process such as postmortems, check for existence and frequency.

4. Deviations from established targets and action limits are recorded and analyzed, and corrective action is taken; or look for rigor of application.

   Discussion: This will be straightforward for quality checkpoints (Fig. 5.19 and Table 5.5). Look for rigor of application, analysis, and corrective action. Especially look for an out of control report for deviations. For other processes, review the performance measures—target versus actual—and corrective action for deviations.

   Here are some powerful questions to ask and steps to take when process deviations are detected. Deviations will be found in the quality checkpoints charts in manufacturing, the sales process, postmortems, and just about every process: When did you last see this problem? If it occurred before, probe as follows:

   - What was the fix to the problem?
   - Was it a good fix?
   - Why has the problem recurred?
   - Review the previous corrective action or out of control report. Does it exist? Is it complete? Look for a good PDCA cycle for every situation.

5. The process has been routinely improved and documented, including the update of standards. Future improvements are evaluated.

   Discussion: It is usually not possible to improve every process, every year. Hence look for past, current, or future plans. Key processes should be competitively benchmarked—probe for this.

Processes to select for the review. Processes to select will include those we have mentioned in Chap. 5; at a design and manufacturing-type entity they include

- In marketing: Product definition and competitive benchmarking
- In R&D: Product definition, product design for reliability, design standards, and product postmortem

Note: In a postmortem review at a manufacturing-type entity, we suggest you review one problematic product. This will allow tremendous insight and analysis of the entity's quality effort, improvements, and ability to learn from a difficult situation.

- In manufacturing: The quality assurance system (including quality checkpoints), supplier management, and product delivery

At a sales and marketing-type entity the processes include:

- Sales and marketing: Sales process management, product and service delivery, and sales won or lost postmortem

Note: In a sales won or lost postmortem, we suggest you review one successful product (with high market acceptance) and one problematic product (with low market acceptance).

- After-sales support or service: Delivery (or the order fulfillment cycle from customer order to customer delivery), installation, repair, and other after-sales activity

Conducting TQC Audits, or Reviews   263

Daily process management and standards scoring. This page is set aside for scoring the items reviewed. Use the scoring matrix below and questions provided for determining the scores. Space is provided to score four processes.

	Score	Average score
1. Is the concept of standards well understood?		_____
2. The key process is identified, documented, and understood.	1.____ 2.____ 3.____ 4.____	_____
3. Key process is monitored (or implemented) and performance against targets and action limits is checked routinely.	1.____ 2.____ 3.____ 4.____	_____
4. Deviations from established targets and action limits are recorded and analyzed, and corrective action is taken; or look for rigor of application.	1.____ 2.____ 3.____ 4.____	_____
5. The process has been routinely improved and documented and standards have been updated. Future improvements are planned.	1.____ 2.____ 3.____ 4.____	_____

Total score: (Maximum 20 points) _____
Rescale: Total score /5   (Maximum 4 points)   _____

Not using data or quality management tools. No evidence of effort. No results or evidence of quality.	Using some data and quality management tools. Evidence of effort in a few areas. Numerous opportunities exist. Little success, lots of opportunity.	Knowledgable use of data and quality management tools. Evidence of effort in several areas. Further deployment possible. Some success, but more needed.	Good use of data and quality management tools. Some innovative approaches. Evidence of effort and deployment in most areas. Successful in most areas with some areas requiring more attention.	Excellent use of data and quality management tools. Many innovative approaches. Evidence of effort and deployment in all areas. Good to excellent results in all areas.
Score: 0	1	2	3	4

Scoring matrix.

Total participation checklist. In this section, we review employee participation activities and management leadership in guiding and sustaining the TQC effort.

1. The quality effort is effectively managed via a strong leadership.

    Discussion: Check for leadership in generating the quality vision, with a specific goal and supporting strategies; communication of the vision and quality activities; and progress toward the vision. Included here will be a quality steering committee led by senior management.

2. Project or quality team activity is well managed.

    Discussion: Check for project linkage to annual plans, key processes, and customer issues as well as for overall coordination and results. Look for project teams, quality teams, and MK, or management-led, teams.

3. Systems and processes are in place to capture and use employee suggestions and to recognize employee contributions.

    Discussion: For employee suggestion schemes, review management of the system. Also look for other means of employee recognition, for example, recognition for contributing patents or special customer satisfaction efforts.

4. Management reviews and recognizes project and quality teams and employee contributions.

    Discussion: Check for regular quality circle conventions and awards to employees for their contributions. In both cases, senior management should be involved.

5. TQC and other education programs are defined, implemented, and sustained.

    Discussion: Review education program for completeness, coverage of all employees, and proper implementation and ensure program is sustained.

Total participation scoring. This page is set aside for scoring the items reviewed. Use the scoring matrix below and questions provided for determining the scores.

	Score
1. The quality effort is effectively managed via strong leadership.	_____
2. Project and quality team activity is effectively managed.	_____
3. Systems and processes are in place to capture and use employee suggestions and to recognize employee contributions.	_____
4. Management reviews and recognizes project and quality teams and employee contributions.	_____
5. TQC and other education programs are defined, implemented, and sustained.	_____
Total score: (Maximum 20 points)	_____
Rescale: Total score/5 (Maximum 4 points)	_____

Not using data or quality management tools. No evidence of effort. No results or evidence of quality.	Using some data and quality management tools. Evidence of effort in a few areas. Numerous opportunities exist. Little success, lots of opportunity.	Knowledgable use of data and quality management tools. Evidence of effort in several areas. Further deployment possible. Some success, but more needed.	Good use of data and quality management tools. Some innovative approaches. Evidence of effort and deployment in most areas. Successful in most areas with some areas requiring more attention.	Excellent use of data and quality management tools. Many innovative approaches. Evidence of effort and deployment in all areas. Good to excellent results in all areas.
Score: 0	1	2	3	4

Scoring matrix.

The TQC Review score can be normalized to 100% and then compared against the comments listed below to determine overall maturity.

Enter rescaled scores for each element, from the detailed checklist.

Then normalize per the instructions given below.

Customer Obsession _____ + Planning Process_____ + Improvement Cycle _____ +

Daily Process Management ___ +Total Participation_____ = Total_____

***TOTAL_____ X 5 =_____% , **Enter this number below to get comments.**

Percentage Score	Percentage score of entity reviewed	Comments on maturity level
0 - 20%		Little knowledge and little evidence of effort in any area. Virtually no attention to quality. Management provides no leadership for the quality effort. Potential of failure in the marketplace is high.
21 - 40%		Some evidence of effort in a few areas. Poor integration of efforts. Largely based on reaction to problems with little preventive effort.
41 - 60%		Some successes. Evidence of effort in many areas and good in some. A mixture of reactive and preventive processes, but many areas lack maturity. Further deployment and results needed to ensure continuity across the entity.
61 - 80%		Good level of success. Evidence of effective effort in most areas and outstanding in several. Good prevention based processes in place with some requiring more attention. Management provides good leadership.
81 - 100%		High level of success. Evidence of effort in all areas and outstanding in most. Full deployment with good integration of processes. Mangement provides strong leadership in the quality effort. Potential to be a world leader.

Figure 8.3 Scoring comments.

Conducting TQC Audits, or Reviews   267

The review team

The review team can consist of two managers. These managers should be experts in quality—they should be trained in quality methodologies, such as those mentioned in Chap. 6. And certainly, at a minimum, they should be familiar with everything discussed in this text. They should be mature managers, and it would be ideal if one of them were a general or a quality manager. It is important that you develop this type of expertise within your organization. The organization's quality manager must organize this activity.

Review techniques. Review team members can start by practicing on several departments per agenda no. 3, which we showed earlier, and then move on to reviewing an entity. During an actual review, the review team should allow the presentations and discussions to flow smoothly. It is important that the review team avoid the temptation of interrupting and proposing a better way—it would be far better to probe at the weakness and allow the person or manager reviewed to understand and admit their weakness. According to W. J. Harmayer in *Audit Interview Technique:*

> It is not a [reviewer's] function to make instant evaluation and comment during the interview. If you suddenly ascertain that these people haven't yet discovered the wheel, make haste slowly. There might be something better than the wheel for the job at hand! If you have, in fact, unearthed the crime of the century, it will keep. Don't rush to advertise it.

Hence, during a review, the review team would first look for the item on the checklist. If present, the reviewers note it down and move on. If not present, the reviewers probe for it and ask questions and move several layers down. Let us review some examples:

- You are reviewing the section on customer obsession, question 2. You have seen the results of a customer satisfaction survey. The next thing you look for is a corrective action plan, that is, what action will be taken to resolve the customer issues. You would then look at the actual implementation of the corrective action plan and the feedback to the customer. After this, you can move to the next question.

- You are reviewing a project in the section on the improvement cycle; you would typically wait until the presentation is completed, before asking any probing questions. The following discussion explains why.

Refer to question 1 in the improvement cycle. During the presentation, you could have been satisfied with the problem definition and schedule of activities. But you had some concerns with the goal—maybe it was not aggressive. You wish to probe deeper, and you want to recommend the rule of 50 percent reduction in defects that we discussed earlier—but you would do this after you see the results and understand how the reviewees set their goal. Maybe they were not aggressive and discovered that they have far exceeded the goal. Now you can convince them that they need to set aggressive goals and propose the 50 percent rule. Alternatively, they may have set too aggressive a goal of, say, 80 percent defect reduction. If they fail to achieve this, you again have a discussion point.

You would take the same tack if the reviewees had no PDCA schedule of activities—resulting in an open-ended project or no goal or no verification of causes from a cause and effect diagram.

To reiterate, the approach a reviewer takes is to avoid the temptation of jumping in with advice at the first opportunity. Rather, the reviewer must let the discussions flow and must probe at weaknesses. If this is done properly, the reviewees will understand their weaknesses. At that point, the reviewer can make recommendations for improvement. *Remember, someone who understands his or her weakness will be more willing to learn, improve, and accept the review team's recommendations.*

Polite but probing questions. The reviewer needs to be polite and firm. Discussions should stay on course and focus on the process and then on the content. The time shown in the schedule is tight but will be sufficient.

The reviewers need to ensure that there are no dog and pony shows. The discussions need to be focused and should stay within the review checklist and agenda. Listed below are questions that can help the reviewers control the flow of discussions and extract information quickly but politely:

- We have a very tight agenda, so we need to move on . . .
- Were you able to determine the root cause of the problem?
- How will you share what you have learned?
- How do you convey your vision (or plans)?
- You are doing a great job responding to customer problems and concerns. How do you link these to future plans and products?
- What would it take to have 100 percent of your customers satisfied or for your customer to become a reference site for potential customers?

Preparing and issuing the TQC review recommendations

Immediately after the review, we suggest that the review team give a verbal report. In the agendas that we provided earlier, we left about an hour for the review team to prepare a verbal report. The verbal report will consist of:

- A discussion on the strengths and weaknesses of each TQC element and recommendations for improvement.
- Five key recommendations—preferably one from each TQC element.

We suggest giving a summary of key recommendations because, typically during a review, you may uncover numerous weaknesses. But you need to prioritize recommendations into just four or five items; this ensures that the entity being reviewed has a manageable list of recommendations.

A written report on the TQC review should be prepared and issued within about 3 weeks. The written report should be formatted as follows:

- A summary of the key recommendations. Keep the list short, with a maximum of five items
- A detailed list of strengths and weaknesses for each of the five TQC elements
- Examples of recommended best practices—similar to those provided in this text

A sample TQC review report is shown in Fig. 8.4. In it, we follow the proposed format discussed above.

Follow-up corrective action. After the report is received, the reviewed entity prepares and implements a corrective action plan. The plan should, preferably, be incorporated into the entity's annual plan. Successful implementation will result in a stronger and more competitive organization.

Common problems discovered during the review process and some review guidelines

What are some of the common problems or weaknesses that you will find during a review? We give you a short list below. In addition, we give you some guidelines on focusing your efforts if you are conducting a review.

Page 1 of 10

Date: December 5, 1990
Subject: TQC Review

To: The General Manager
XYD Division

We conducted a TQC Review at XYD Division on November 26 and 27. Listed below are the key recommendations. In addition, we have attached a detailed report, giving your strengths, weaknesses and opportunities for improvement. A scoring summary is also attached.

Page 2 of 10

Page 3 of 10

2. **PLANNING PROCESS**

   Strengths observed:
   The General Manager and his staff had clear plans that linked and cascaded well down the organization.

   Opportunities for improvement:
   Item 2a:
   Several objectives – at GM and lower levels – had no goals. This makes it very difficult for you to measure progress toward your plans. Additionally, the lack of a measurable goal has made it difficult for you to generate suitable strategies or methods for achieving the objectives.

   Recommendation 2a:
   Provide measurable goals for your objectives. For example, your objective of "Increase Productivity", should have a specific goal of, say, 10%. Supporting strategies should be better formulated and add up to your goal. In the appendix, we provide a short discussion on generating strategies.

   Item 2b:
   Etc.

Figure 8.4a A sample of a TQC review report.

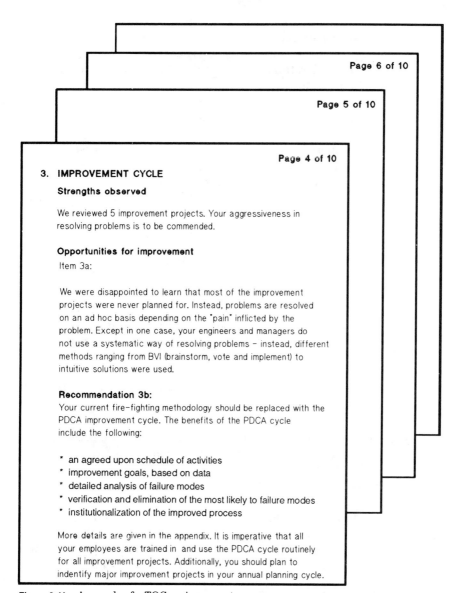

Figure 8.4b A sample of a TQC review report.

1. Reviewing Customer Obsession
   a. Review the entity's approach to measure and improve customer satisfaction and competitiveness.
   b. Look for understanding of customer needs, concerns, and follow-up in the areas of:
      (1) Customer satisfaction surveys
      (2) Customer complaints and feedback management
      (3) Customer meetings and visits
      (4) Competitive data, benchmarking, and sales won or lost analysis and reporting
   c. Common problems or weaknesses observed:
      (1) Weak system to collect and collate customer and competitor data
      (2) Insufficient follow-up or corrective action after getting the data
      (3) Lack of or weak sales won or lost analysis
   d. During the review, focus on:
      (1) The analysis of data from the various systems such as customer surveys, customer complaints, competitive benchmarking, and sales won or lost analysis
      (2) Follow-up activity and corrective action as a result of the analysis
      (3) Select two or three items (one from customer complaints, one from the survey, etc.) and follow it all the way through resolution
2. Reviewing the planning process
   a. The general manager and his or her staff should present their plans and discuss progress to their plans.
   b. Look for:
      (1) A standardized planning process, including:
         (a) A process to generate key issues
         (b) Influenced by customer, quality, and delivery issues.
         (c) Separation of plans into Hoshin or breakthrough type objectives and routine Business Fundamentals type of activities
         (d) The annual (Hoshin type) objectives and detailed plementation plans
         (e) Business Fundamentals plans.
      (2) Robust supporting strategies
      (3) Proper cascading of plans down the organization

            (4) Compact and concise plans
            (5) Regular reviews of plans
        c. Common problems and weaknesses observed:
            (1) Poor targets and metrics
            (2) Too many objectives—attempting to do too much
            (3) Weak or no detailed implementation plans for each strategy
            (4) Nonrigorous reviews or, worse still, no reviews of progress of plans
            (5) Corrective action or management intervention when it is apparent that plans will not be achieved
        d. During the review, first look at the process, then the content. Be prepared to discuss strategies for robustness and their contribution to objectives.
3. Review of the improvement cycle
    a. Select three or four projects linked to the annual plan or crucial issues.
    b. Look for a systematic problem-solving process:
        (1) Go through the presentation
        (2) Note if sequence follows PDCA standard
        (3) Discuss any deviation from the PDCA approach to problem solving
        (4) Discuss the aggressiveness of improvement targets
    c. Common problems and weaknesses observed:
        (1) Approach to problem solving is not systematic:
            (a) Fire fighting or genius/intuitive approach
            (b) No schedule or goal for project
            (c) New improved process not standardized; the problem may recur
    d. During the review focus on:
        (1) The problem-solving process
        (2) Understanding of the PDCA improvement methodology
4. Review of daily process management
    a. Review standards and processes in recommended list or for a unique entity select their key processes. This may require some discussion.
    b. Look for process thinking:
        (1) The concept of standards
        (2) Documentation
        (3) Monitoring of the process with performance measures
        (4) Analysis and control of deviations

c. Common problems and weaknesses observed:
   (1) Lack of standards
   (2) Poorly documented or loosely managed processes
   (3) No process performance measures and targets
   (4) No or weak analysis of deviations in processes
   (5) Weak corrective action when deviation occurs
d. During the review focus on:
   (1) Benefits of each process
   (2) Documentation of process
   (3) Good management of process
   (4) Improvement plans (past, present, or future) for each process

5. Total participation
   a. Discuss with general or quality manager.
   b. Look for total participation:
      (1) Management leadership in quality
      (2) Entity quality direction
      (3) Quality circle, or team, activity
      (4) Employee recognition and suggestion schemes
      (5) Training program
      (6) Management involvement
   c. Common problems and weaknesses observed:
      (1) Lack of management leadership in setting and managing the overall entity quality direction
      (2) Weak management of quality circle, or team, activity
      (3) Insufficient support from quality department
   d. During the review focus on leadership via management and active participation of these activities

Presidential Audits, or Executive Reviews

The TQC review process we have proposed is fairly robust and can be very effective in probing for weaknesses and improving the TQC effort in an organization.

In addition to the TQC review, we recommend a presidential audit, or executive review. Presidential audits are common in some large Japanese companies, and we know of at least one American company—Florida Power & Light—that practices it. The presidential audit can be a very effective way for a president or chief executive of a company to review a product or sales division of any large operation.

The objective of a presidential audit, or executive review, is very similar to that of a TQC review. In addition, this review allows the president or chief executive to convey his or her commitment, support, and leadership in the TQC effort. This is a powerful method for the chief executive, and it allows him or her to understand the organization's situation better and to steer the quality effort. It will, however, tend to be a shorter and more focused review. This is obvious from the agenda that follows.

Review agenda for presidential audit

The review can be structured in many different ways. Since time and resources are limited, we suggest the following:

What is reviewed	How long (hours)
1. Review of customer obsession	2
2. Review of planning process	2
3. Review of improvement cycle	2

As in the case of the TQC review, this audit should focus on the process and then on the results—it should not be a dog and pony show. The audit can be completed in about 4 to 6 hours. Such a focused agenda will allow the president or chief executive to address other topics in a busy day.

We now provide you a review checklist for a presidential audit. It consists of a list of key questions, followed by a discussion that sets expectations.

Customer-obsession checklist for presidential audit. The purpose here is to ensure that the entity being reviewed pays a great deal of attention to its customer needs and is staying close to them. The key items to check for include follow-up on customer surveys, customer inputs, resolution of customer complaints, and introduction of new products and services that will meet current and future customer needs.

1. Customer needs and satisfaction levels determined and follow-up corrective action taken on products and services.
   Discussion: Here we look at the process of determining customer needs and satisfaction levels, an analysis of the available data, and

how this analysis influences current products and services and future plans. Items to look for include surveys, customer focus groups, and customer feedback via surveys and response cards after purchase.

2. Customer interface, visits, and follow-up activities are well established.

   Discussion: Here we determine methods of visiting and meeting with customers to get their inputs, concerns, and suggestions; what is done with the data; and how analysis of this data influences current products and services and future plans.

3. Product and service quality data collection processes are well established. Data is used to drive improvement.

   Discussion: Here we look at the process of collecting failure data on products and services from customers after the sale and how analysis of this data influences current products and services and future plans. Items to look for include product failure information and the customer complaint and feedback system.

4. Competitive data is available and analyzed; the analysis is used to influence both current and new products and services. This will include a sales won or lost analysis and follow-up, corrective, action.

   Discussion: Here we look at competitive benchmarking and other means to get competitive data. For a sales and marketing department, we would look at a sales won or lost analysis, including an understanding of why sales are won or lost and follow-up corrective action.

Note: For all the above items, it is imperative that we find out what is done with the collected data.

Planning process checklist for presidential audit. This section allows the president or chief executive to determine if his or her objectives or direction have been picked up by the entity being reviewed and if they are being translated into implementation plans at some lower level. Additionally, progress of these plans is checked.

1. A planning process, with the following features, is in place:
   - Long-term and annual plans.
   - Generation of key issues, and from them a few breakthrough objectives.
   - Influenced by previous year's issues.

- Quality, cost, education, and delivery issues have been addressed and improvements are planned; customer issues are addressed and linked to issues from the customer-obsession checklist. If not addressed, good analytical explanation is available.

    Discussion: Plans are presented and reviewed; check for the features mentioned above. Remember to check for nonachievement of previous year's objectives and lessons learned.

2. Plans are robust and include the following:
    - A few breakthrough objectives, with robust supporting strategies.
    - Goals and performance measures can be measured.
    - Higher-level plans are properly deployed down the organization.
    - Cross-functional issues are addressed.

    Discussion: When plans are reviewed, focus on the one or two objectives considered the most crucial, then check for the above features throughout the plans; that is, check for breakthrough objectives with appropriate strategies, measurable goals and strategies, deployment down the organization, and cross-functional team activity. Check how the cross-functional teams are monitored and managed. Also, be especially wary if management has too many objectives—plans must be specific and exclude 100-year objectives.

3. A robust Business Fundamentals plan exists and includes regular tracking of performance.

    Discussion: Every department must have such a plan. Remember to ensure that the department reviewed distinguishes breakthrough (Hoshin) from Daily Management issues.

4. All plans are reviewed regularly for progress and corrective action taken, including changes to the plan when necessary.

    Discussion: Check the review process, progress to plan, lessons learned, and follow-up corrective action.

Improvement cycle checklist for presidential audit. At least one improvement project that is linked to the annual plan of the entity should be reviewed. Where possible the project should be linked to, or originated from, the president's or chief executive's objective—for example, a request to cut costs or to improve product and service quality.

1. Problem definition and linkage to chief executive's plan or a crucial issue.

Discussion: A project supporting the president's or chief executive's objective should be selected. Alternately the project should be driven by the annual plan or crucial issues. Problems, definition, processes, improvement goals, and project schedules are defined and available.

2. Data collection, analysis, and understanding of root cause.

    Discussion: Good collection of data and appropriate use of quality tools such as Pareto and cause and effect diagrams and graphs. Most likely causes have been identified and verified with data.

3. Alternative solutions are reviewed and results achieved. Deviations from goals are understood.

    Discussion: Alternative solutions are reviewed, evaluated, and implemented. Results have been achieved and measured; deviations from goals are well explained and acted upon.

4. Review standardization and future plans. New process is standardized.

Discussed: New or modified processes are documented and employees are trained; future plans exist. Look for a method to update standards across the organization, for example, a method similar to the SUR format, discussed earlier. *Most important, look for lessons learned and reasonable assurance that this problem will not recur at the entire entity—you must probe extensively for this.*

Issuing recommendations after a presidential audit, or executive review

Recommendations, concerns, or issues that are raised at a presidential audit, or executive review, should be conveyed during or immediately after the review. One of the objectives of such a review is to ensure that the president or chief executive gets the opportunity to convey his or her commitment, support, and leadership in the TQC effort. Hence, immediate, verbal feedback is essential. The review techniques are similar to our earlier discussions.

Summary: Conducting TQC Reviews

The TQC audit, or review, is a very powerful tool to maintain or increase the momentum of a TQC effort. It can help uncover weaknesses, ensure that managers and employees understand the weaknesses, and provide opportunities for improvement, which can increase business success.

We also recommend the use of presidential audits, or executive reviews, as a means to ensure the chief executive's support, commitment, and leadership in the quality effort. In the final analysis, the quality effort is directed by the chief executive, and such reviews allow him or her to understand and direct the effort.

The TQC review and presidential audit are among the best methods that you can select to ensure a strong, competitive, and more successful organization.

Chapter 9

Some Thoughts on the Essence of TQC

To return to the root is repose; it is called going back to one's destiny. Going back to one's destiny is to find the Eternal Law.
Tao Teh Chin (The Way)
LAOTSE

What is the essence of TQC? By now, you would have guessed that it includes a focus on the process, or "the way," rather than only on achieving results. Kaoru Ishikawa called it thought revolution in management; and that is what TQC is—a better way to manage. We offer a few concepts which get to the heart, the very essence, of TQC.

Quality First

Quality comes before anything else; there are several items of importance here.

Continuous improvement

Continuous improvement is a fundamental concept of TQC—we strive for continuous improvement and aim for perfection. The concept of continuous improvement is predicated on the knowledge that breakthroughs will be few and far between—hence incremental improvements must be encouraged in all processes, products, and services.

Often, there is a concern about investing in continuous improvement—because a company feels that it is good enough or already competitive or that there is balance between cost and quality. This complacency is dangerous in any organization because complacency is the very enemy of quality. This is succinctly expressed as the whispering of Satan by Takoshi Hokake:

> "...balance between cost and quality" is such a phrase that affects us like opium. Tearing off this veil of this beautiful phrase "balance of cost and quality"—the "quality first" policy should be adopted. This is the very energy source. But I have no weapon to shut out this whispering of Satan....

Hence, we stress the need for driving toward continuous improvement. But when problems or defects run at a very low level, it is time to look at the *second dimension of quality*—the area of attractive quality.

Attractive quality

We discussed this in detail in Chaps. 1 and 2. In the spirit of continuous improvement, when problems and defect levels are low, we must start providing more attractive quality items. An example of this is found in the automobile industry. Initially, Japanese manufacturers kept ahead of American and European manufacturers by providing high quality and (almost) defect-free cars. By 1990, the American and Europeans were catching up. But the Japanese have leapt ahead by going beyond providing defect-free cars, by focusing on attractive quality items—items such as ergonomics, extraordinary *touch and feel* features, superior service, and so on. In Japanese, this attractive quality feature is called *miryokuteki hinshitsu*. This attractive quality is what we define as the second dimension of quality. *If we are routinely operating in this second dimension of quality, we will be operating in a new zone of TQC—total quality creation. Clearly, this is where we want to be. Always.*

Supporting methodologies and techniques

Some of the supporting methodologies and techniques include the PDCA cycle, quality function deployment, customer complaint and feedback system, customer surveys, competitive benchmarking, postmortems, employee suggestion schemes, and quality circles, or teams.

Customers First

Customers always come first. They pay our wages and bills and help us to be successful. "The next process is our customer" is an extension of the customers first concept. In an organization, everyone is part of a process and everyone has a customer. That customer must also be satisfied, not only the external customer. This creates a chain of suppliers satisfying each customer until we reach the end customer.

Supporting methodologies and techniques

Some of the supporting methodologies and techniques include: customer complaint and feedback system, customer surveys, competitive benchmarking, postmortems, and quality function deployment.

The Importance of the Way, or Process

Getting results is very important, but they should come as an outcome of a good process. This will ensure repeatable and predictable results, which requires that management and employees focus on using standards. An example is using standard methodologies for design rules, planning (Hoshin), improvements (PDCA), and other processes. These include processes such as the quality assurance systems, the sales process, and the postmortem process.

When experienced managers or employees leave, they are irreplaceable. But the loss is less severe in a process-oriented organization, in which we capture their experience and best practices into standards at our organization. In addition, well-managed processes can be a leading indicator of good results in a company. The outcome is less dependence on the genius of individuals and more dependence on the system, the way, or established processes to get results.

Balance between standards and creativity

In implementing standards, there must be a balance between using standards and having time to be creative. Often, the argument against using standards—such as formal planning or the PDCA improvement cycle—is that they stifle creativity. In TQC philosophy we emphasize the use of proven processes or standards. We save our creative energies for new unexplored areas. This is how we get maximum output and efficiency in an organization.

Supporting methodologies and techniques

Some of the supporting methodologies and techniques for managing processes are the PDCA cycle, competitive benchmarking, customer complaint and feedback system, and daily process management—including management of key processes and postmortems.

The Organization That Learns and Grows

It is crucial that the organization learns and grows. This implies a willingness to take risks, to learn from mistakes and successes, to train continuously, and to drive change in the organization.

Organizational learning is extremely important. This includes learning from mistakes and the lessons of history. Whenever an error or problem is encountered, we do not just fix it and move on. Instead, we must do the following:

- Understand why the problem was created.
- Understand why the problem was not caught at creation.
- Ensure that the problem will never occur again in our department.
- Ensure that the problem will never recur again in our entire organization, in all our future products and services. This is very important.

Another unique way to ensure learning is via cross-functional task forces, that is, task forces consisting of senior managers that form a team which analyzes and resolves issues that cross functional boundaries. Finally, best practices that occur in one part of the organization must be quickly propagated and emulated throughout the rest of the organization.

Supporting methodologies and techniques

Some of the supporting methodologies and techniques include the PDCA cycle, postmortems and T-type matrix, Hoshin plan reviews, cross-functional task forces, out of control reports, CAR and SUR formats, education, and TQC reviews.

TQC: Organization and People Running at Peak Efficiency

The outcome of being strong in the essence of TQC is to have an organization and people operating at peak efficiency. A good analogy is to look at the automobile industry. General Motors' market share continues to decline, while Toyota's continues to increase. General Motors continues to invest heavily in automation and robotics, but the money seems to disappear into a black hole. Toyota, with less capital investment per employee, is able to have higher productivity and operate more effectively. Its average time to design a car is 3 years against 5 at American manufacturers; the Toyota-Lexus luxury car is assembled in one-sixth the time that Mercedes takes. Toyota has succeeded partly because of its TQC effort, which includes its continuous improvement philosophy and its world-famous just in time manufacturing system. Toyota and some of its subsidiaries have won the Deming Prize for their quality and customer focus. Toyota's operations and people run at the highest efficiency in the automobile industry.

TQC, Creativity, and Success

Does TQC methodology ensure business success? To answer this question we need to review the attributes of a successful company. The following is a list of important attributes:

- Keeping close to the customer
- Good processes
- Good product and service strategies
- Stable organization
- Strong leadership

TQC methodology helps to provide many of the above attributes. It certainly ensures keeping close to the customer, it provides good processes, and it influences good product and service strategies. These three attributes provide a good basic system. The other two attributes of a stable organization and strong leadership must be provided by management.

With these basics in place there is stability; this provides ample opportunity for creativity in new areas, in beating the competition, and in planning for the future. The combination of all these provides a strong, dynamic, competitive, and successful company.

Conclusion

The necessity and benefits of TQC have been extolled throughout this book. By the time you get to this point, a good understanding of TQC concepts, techniques, and methodologies has been conveyed. You are definitely ready to start implementing TQC. If you have already started, you will be able to improve and accelerate your current effort.

A vital point to remember is that the goal is one of continuous quality improvement. TQC is only a philosophy, a tool, a means. Remember, *TQC is a journey not the destination.* Once started, you will be on an unending journey toward better quality. With a quality focus, you will achieve increased sales, increased productivity, and increased profits. Quality is the key to survival, success, prosperity, and happiness.

Appendix

1

Improvement Project to Reduce Dissatisfied Customers Using the PDCA Cycle

Plan Stage

Project Theme

During the recent customer satisfaction surveys, our overall satisfaction score was quite good, at 7.4 on a scale of 1 (worst) to 10 (best). We scored poorly in the category of sales interactions. This is an area of concern since this means that our customers are dissatisfied with how our company and sales representatives interface with them.

Our project theme is "to reduce customer dissatisfaction in the area of sales interactions." The target will be set after analysis of the data.

Project schedule. The following project schedule was agreed upon:

Activity	Timing
Select project theme and prepare schedule of activities	January
Grasp the present status and set target	January
Analyze the cause and determine corrective action	February to May
Implement the plan	June to October
Check the results via another survey	March
Standardize the results	April, May

Grasp current status

In the customer satisfaction survey there are 10 categories. The results of the survey are shown in Fig. A1.1. The figure shows that sales interaction and cost of ownership were the two lowest-scoring categories.

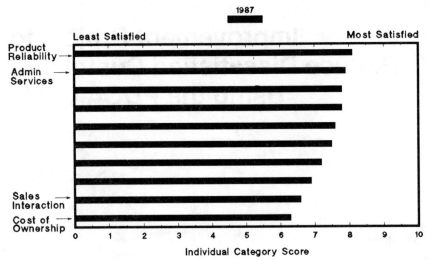

Figure A1.1 Customer satisfaction by categories.

We focused in on sales interaction since we felt we had an opportunity here to improve; cost of ownership was also very important, but it was passed over since there was already a separate project to reduce costs in the company. Sales interaction refers to how well our sales representatives relate to and interface with our customers.

We were able to stratify the sales interaction data by satisfaction levels. This is shown in Fig. A1.2. From Fig. A1.2, it is evident that 33 percent of our customers were dissatisfied with sales interactions, that is, they were unhappy with the way our company or sales representative interfaced with them. We set our project theme and target as follows: To reduce dissatisfied customers in the sales interaction category by 50 percent.

Our analysis continued and we reviewed the survey forms that each customer had completed. We noted customer comments for the sales interaction category. Comments ranged from "I am very happy with my sales representative" to "After the sale is made your sales representative disappears."

The sales interaction category had seven subcategories, and they were each rated in the survey on a scale of 1 to 10. Data was available for each of the subcategories in terms of percentage of dissatisfied customers, that is, those rating us 1 to 5 (the need improvement or dissatisfied scores). This data is shown in the Pareto diagram in Fig. A1.3. *Note:* The data shows percentage dissatisfied in each category.

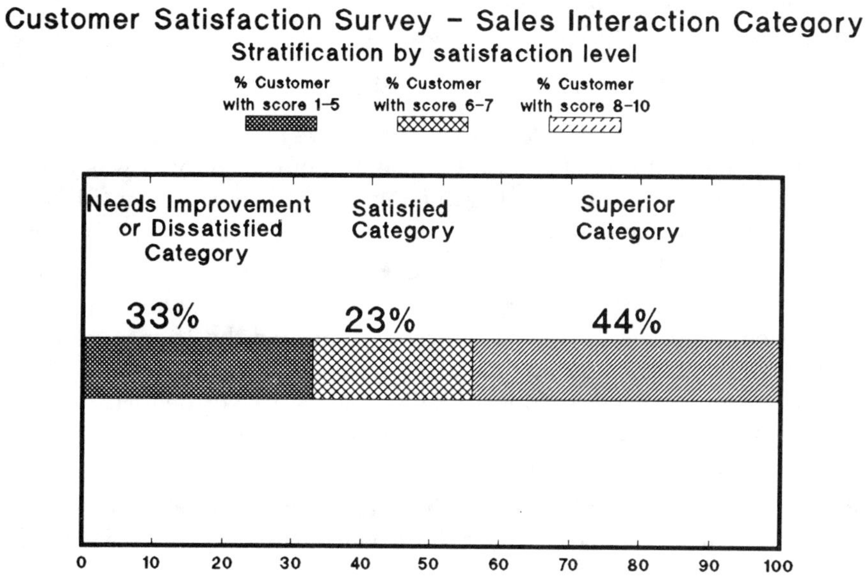

Customer Satisfaction Survey – Sales Interaction Category
Stratification by satisfaction level

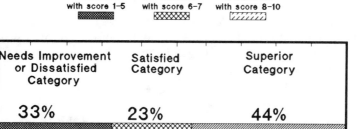

Figure A1.2 Customer satisfaction for sales interaction.

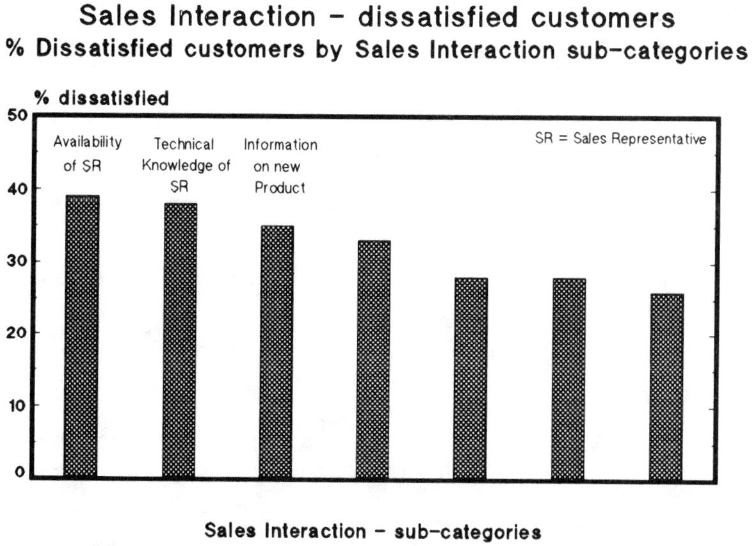

Sales Interaction – dissatisfied customers
% Dissatisfied customers by Sales Interaction sub-categories

Figure A1.3 Dissatisfied customers for sales interaction.

Analyze the cause and determine corrective action

This was separated into three steps.

1. *Prepare cause and effect diagram.* Here we determined the causes for customer dissatisfaction. A cause and effect analysis was done for the first three Pareto bars of Fig. A1.3. This is shown in Fig. A1.4. Our purpose was to understand the most likely causes and eliminate them.
2. *Prepare hypothesis and verify most likely causes.* Our team reviewed the various possible causes and selected the 10 most probable ones by voting, based on our experience. These are listed in Fig. A1.5. From this list, the most likely causes were verified with separate independent data. This is also shown in Fig. A1.5. It can be seen that several probable causes were discarded.
3. *Determine corrective action.* The corrective action, shown in Fig. A1.6, was agreed upon. Where there was no direct corrective action, we proposed an alternative.

Do Stage

Implement corrective action

The proposed corrective action was implemented over a period of 5 months.

Check Stage

Check the results

Another survey was held 16 months later. The following effects were observed:

- The overall survey rating was 8.22, an improvement over the previous survey.
- The rating for sales interactions was 7.03, also an improvement over the previous survey.
- The percentage of dissatisfied customers in the sales interactions category reduced dramatically, as shown in Fig. A1.7.

Figure A1.7 shows that we met our goal of a 50 percent reduction in dissatisfied customers in the sales interaction category. A marked im-

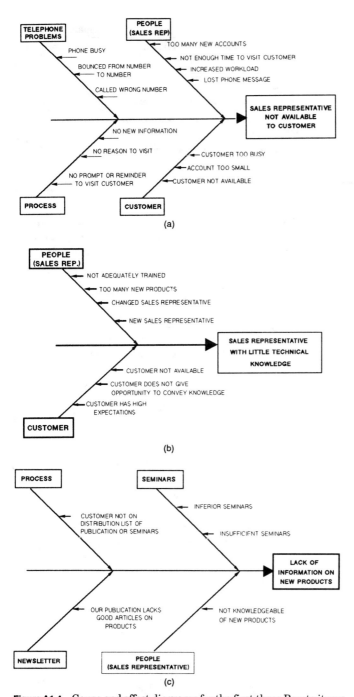

Figure A1.4 Cause and effect diagrams for the first three Pareto items.

Appendix One

	PROBABLE CAUSE (Hypothesis)	VERIFICATION OF MOST LIKELY CAUSE	REMARKS	
1a	Increased workload	Compared to last year SR has more workload due to: (a) Collection of sales process data (b) Implementation of Product support plan for every proposal This has caused about 4 hours or 10% more work per week	This is an issue of concern.	✓
1b	Too many new accounts	Overall new accounts have increased by 8% but 5% of old accounts have dropped and overall change is 3%	This is not a major cause	✗
1c	No prompt or reminder to visit a customer	This is correct. Currently there is no system to remind SR who to visit and when	This is an issue	✓
1d	No reason or new information, hence why visit customer	Data does not support this statement. Last year we introduced over a 100 new products. Also we held 7 seminars.	This is not an issue. There are good reasons to visit customers.	✗
1e	Telephone problems o Phone busy o Never returned call o Called wrong number o Bounced from number to number	We were not sure. We checked with 30 dissatisfied customers via a phone survey: Every customer had experienced one or more of these phone problems.	A major revamp of the phone system or process is needed. These are perennial problems.	✓
1f	Account too small	40% Sales are of a value less than $20K per year	SR feels that they do not want to visit these accounts.	✓
2a	Too many new products	This contradicts item 1d. There were over 100 new products introduced and 7 new product seminars in the last year.	This is an issue. SR's feel they are not informed and do not understand some of our new offerings.	✓
2b & 3a	Not adequately trained. SR not knowledgeable of new products	Agrees with item 2a. We polled 15 SR's and 80% felt they were inadequately trained	Same as for Item 2a.	✓
2c	New SR not adequately trained	The proliferation of new products and new SR's is causing a training gap.	This is a concern and issue.	✓
3b	Customer not on distribution list of newsletter or seminar	This was difficult to check. So we looked at the dissatisfied customers to see if their company was on the distribution list, but not all customers gave their company name. Of those that did 45% were not on the distribution list.	This is an issue. We lack a good process to keep the list updated, complete and accurate	✓

(Note SR = Sales Representative)

Figure A1.5 Verification of most likely cause.

ITEM		CORRECTIVE ACTION
1a	Increased Workload	o Secretaries to manage paperwork o Automation of data collection and report preparation o Support Administration will provide resources for plan preparation
1c	No prompt or reminder	o SR will prepare a monthly call schedule
1f	Account too small	o Distric Manager will fill in gaps in SR schedule o Not all accounts can be visited, hence other small accounts will be contacted by a Telephone Sales force which will be set up during the next 6 months o Long term plan: Customer data base will be updated based on inputs from telephne sales force and SR visit schedule. Missed visits or calls will be red-flagged every 6 months via exception report to managers.

1e	Telephone Problems	o Revamp phone system as follows:
		  – Provide a single Customer Information center number for all customer needs.
		  – All unanswered calls to SR will get automatically transferred to this number
		  – This proposal will take minumum of 6 months to install

2a	Too many products	We need to increase the knowledge and marketplace competence of Sales Reps via the following: o Conduct monthly, day and night classes on continuous product and sales technique education o Provide more product information in customer newletter which will also be distributed to sales representatives.
2b	Not adequately trained	
3a	SR not knowledgeable of new products	
2c	New SR not adequately trained	Revamp new SR training
3b	Customer not on data base list for newsletter and seminar.	Revamp data base by doing the following: o Update current lists o Ensure monthly updates o No deletions allowed of "inactive" accounts o Add field to indicate visit or call in current year and provide 6 monthly printout if no visit occurs.

(Note SR = Sales Representative)

Figure A1.6 Determine implementation plan.

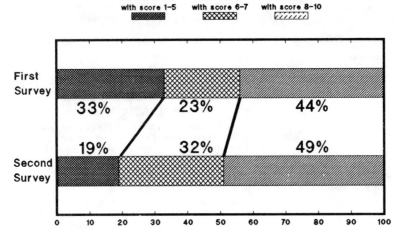

Figure A1.7 Improvement in customer satisfaction for sales interaction category.

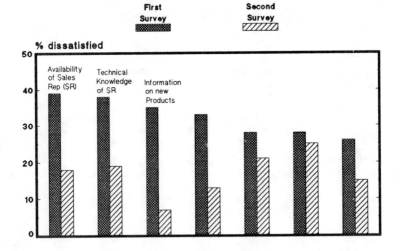

Figure A1.8 Paired Pareto of reduction in dissatisfied customers.

provement was also evident in the subcategories of sales interaction. This is shown in Fig. A1.8. Note the improvement in the first three Pareto bars, where the impact was greatest. This is where we had focused our efforts. There were no unwanted side effects—in fact, improvements were observed in all categories.

Act Stage

Standardization of activity

We will continue all the current activity that was launched after the first survey. In particular, the telephone-sales force and the customer information center will be enhanced; all implemented activities will be documented. Currently more funds are being sought to improve the customer database. We note that training and customer database management are recurring issues; we have proposed a sales process manager job function to manage these on a long-term basis. Our proposal has been accepted.

Future plans

We will now review the second survey data and determine which area to improve.

Appendix 2

The Seven Quality Control Tools and the New Seven Tools

We provide here a very short discussion on the seven tools. This is meant to be an overview. For more information, we suggest you refer to the recommended readings listed in the Bibliography. The discussion is followed by a list of definitions of the new seven quality control tools, which was contributed by Khushroo Shaikh, Hewlett-Packard, Direct Marketing Division. For more details, refer to the Bibliography.

In terms of importance, the seven tools are the most useful. Kaoru Ishikawa has stated that the seven tools can be used to solve 95 percent of all problems. Independent data that we have reviewed supports his statement.

Checksheet

The checksheet is used to facilitate the collection and analysis of data. "Garbage in, garbage out" is an old cliche, but it is true. Therefore, the purpose for which data is being collected must be clear. Data reflect facts, but only if they are properly collected. An example of a checksheet used in the soldering process is shown in Fig. A2.1. The number of defects and the locations where they are found can be recorded and analyzed for causes.

Pareto Diagram

A Pareto diagram is constructed to show the relative importance of different categories in a process, for example, defects, cost, and failure modes. The vital few can be separated from the trivial many. This will serve as a basis for us to select the most crucial items for improvement—typically those at the left-hand side of the Pareto diagram. This will ensure that what we try to improve will have the greatest impact. An example of a Pareto diagram for defective printers is given in Fig. A2.2.

Wave Soldering Defects Record

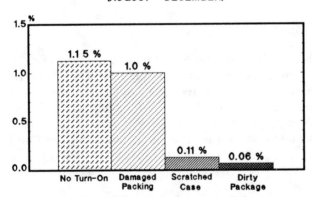

Description	No. of Defects				
	5	10	15	20	25
1. Blow Hole					
2. Pin Hole					
3. Icicycle					
4. Etc					

Total no. of Defects =

$$\text{PPM} = \frac{\text{Total no.}}{\text{No. of Joints}} \times 10^6 =$$

Remarks:

Figure A2.1 A checksheet for soldering process.

DEFECTIVE PRINTERS (AUGUST – DECEMBER)

- No Turn-On: 1.15 %
- Damaged Packing: 1.0 %
- Scratched Case: 0.11 %
- Dirty Package: 0.06 %

Figure A2.2 A Pareto diagram to show major defects.

Cause and Effect Diagram

Causes of problems are numerous. The cause and effect diagram helps us to find out all possible causes, to sort them out, and to organize their interrelationship. Typically, the causes are brainstormed in a free-flowing session; the various causes are then categorized into a few main categories. In the manufacturing environment, the main categories are usually people, machines, materials, and method. The cause and effect diagram is often called a *fish-bone* diagram or an *Ishikawa* diagram, after its creator. Figure A2.3 shows a cause and effect diagram for tasty pizza.

Stratification

Stratification is the technique of analyzing data by separating it into several groups with similar characteristics. It is central to the effective use of the seven tools. Imagine an engineer trying to sort out operator-related problems in his or her production line. The data must be stratified by operators to better understand their individual situation. Since individuals behave differently, specific corrective actions tailored to individuals may be required to improve their performance.

Graph and Histogram

A graph is a pictorial way of summarizing data. It gives us a visual display to reveal the message hidden in a maze of data. Some common types of graphs we use are the line chart, pie chart, bar chart, scatter diagram, and histogram.

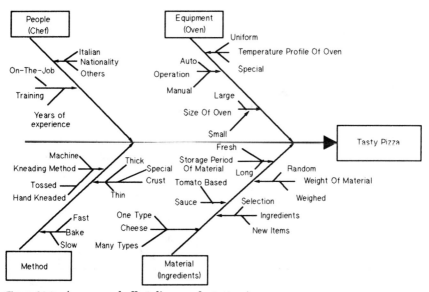

Figure A2.3 A cause and effect diagram for tasty pizza.

A histogram is used to summarize the frequency of occurrence of something from a sample of data. The shape of the distribution is displayed and descriptive statistics can be readily calculated. In the histogram shown in Fig. A2.4, the proportion of items not meeting specification is easily grasped.

A scatter diagram is a special type of graph which shows the relationship between two variables. If there is an empirical relationship between the two variables, it will be seen in a scatter diagram. Furthermore, if the relationship is linear, the correlation coefficient is used to measure the degree of association. In a cause and effect relationship, a regression equation is fitted to describe the behavior. A scatter diagram showing the relationship between labor cost and production volume is shown in Fig. A2.5.

Control Chart

The basic requirement in a manufacturing process is to establish a state of control and to sustain this state through time. Standardization of working methods is necessary to maintain this state. A control chart enables us to observe if this standardization is correct and whether it is being maintained.

A control chart, whether for measurements, attributes, or defects, has a center line corresponding to the average quality at which the process should perform; if statistical control of the process exists, there will be two control limits: the upper and lower limits. The average and

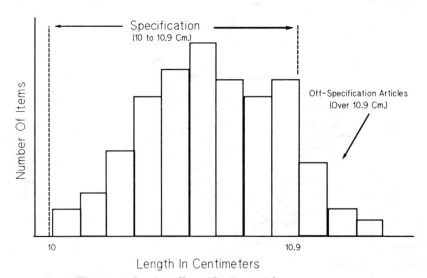

Figure A2.4 Histogram showing off-specification articles.

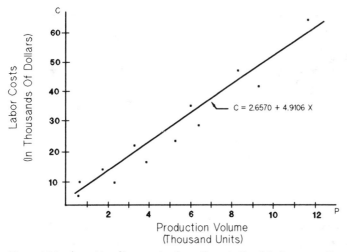

Figure A2.5 A scatter diagram to show the relationship between direct labor cost and production volume.

the two limits are computed from the data—objectively. An out of control situation, trends, cycles, and other unnatural patterns can be easily detected from the chart.

The now-renowned example quoted from Edward Deming's writing can best demonstrate the essence of a control chart. Figures A2.6 and A2.7 show the average scores, x, of a beginning golfer and an experienced golfer.

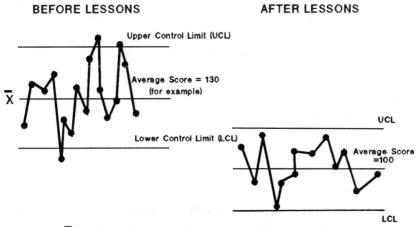

Figure A2.6 $\bar{X}$-chart for a beginner in golf.

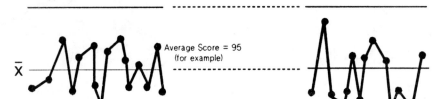

Figure A2.7 $\bar{X}$-chart for an experienced golfer.

In Fig. A2.6, the average scores before lessons were not in a state of control, as evident from the points outside the two control limits, UCL and LCL. There could be many causes: the way the golfer stood, the clubs used, and so on. The causes were identified by a golf professional, and corrective training was given. The purpose of the training was to remove all the identified or assignable causes. After the lessons, the average scores dropped. More consistency was shown since the two control limits were narrower, or closer to each other. The lessons improved the score or system.

In Fig. A2.7, the experienced golfer tried to improve her game by taking lessons. The control charts for before and after lessons showed similar characteristics. The lessons accomplished nothing. Only a new way of training can improve the score or system. For example, a golf professional could do a detailed swing analysis of the experienced golfer's swing and then propose corrective action.

An analogy is easily drawn between the average scores of the golfers and performance of a production process. The inexperienced golfer can be compared to our processes which are not under control. After a proper education and work standardization, we will be able to remove all assignable causes, and thus the process is brought to stability. When the process is in control, as in the case of the experienced golfer, management and engineers must intervene to improve the process—after a thorough analysis.

The New Seven Quality Control Tools, or Seven New Management Tools

The KJ method, or affinity diagram

The KJ method extracts the essence of the message in a large volume of verbal data by using the mutual affinity method. An example is as follows: There could be an enormous amount of verbal data from a customer satisfaction survey or an organizational problem. The numerous verbal comments are put on paper, cards, or Post-it stickers; they are then sorted into categories of issues. The data can then be arranged in a diagram showing the affinity of the issues; insight into solutions can then be determined. This method capitalizes on intuitive reasoning. Kawakita Jiro conceived and developed this tool, hence the name.

A subtle twist to this technique could be to group a large amount of verbal data, for example, complaints and comments, by categories from a customer satisfaction survey; this allows the preparation of a Pareto diagram of categories. Hence, under appropriate circumstances, this methodology allows conversion of verbal data into quantitative data.

The relations diagram, or interrelationship diagram

The relations diagram is helpful in clarifying intertwined relationships in a complex, or multidimensional, situation in order to find an appropriate solution. This tool is useful for understanding complex relationships in a manufacturing or planning process or for introduction of new ideas and programs.

The matrix diagram

The matrix diagram is helpful for understanding multilevel, multidimensional relationships and expressing them in a graphical manner. Applications include deployment of quality in a product (for example, quality function deployment, or QFD), tying goals to methods, and understanding complex relationships between processes and failures (for example, via a T-type matrix).

When we have two variables, we get a two-dimensional matrix which consists of rows and columns. A two-dimensional matrix is a L-type matrix; such a matrix displays relationship information between two variables, for example, goals and methods.

A T-type matrix is obtained when two L-type matrices are combined. In a T-type matrix, variable A has a relationship with variable B. Variable A also has a relationship with variable C, but there is no direct relationship between variables B and C. An example of a T-matrix is given in Chap. 5 in the section on postmortems.

A Y-type matrix is a three-dimensional display of information. There is a simultaneous pairwise relationship between the three variables. In other words, a relationship exists between A and B, B and C, and C and A.

An X-type matrix portrays four variables; the relationships, however, are confined to the adjacent pairs, not to random selection of a pair.

The systematic diagram, tree diagram, or dendogram

The tree diagram is a very systematic approach for identifying means (methods, ways, or approaches) that will be needed to achieve an overall objective or goal. The completed systematic diagram resembles a tree (for example, one main trunk with primary, secondary, and tertiary branches); hence, it is also known as a tree diagram. An organization chart is a good example of a systematic diagram. Another name for this tool is a dendogram. It may be familiar to you if you have traced your family roots.

The systematic diagram helps to identify the starting point for problem solving. It is also useful for developing the necessary implementation plan for a given objective. In certain applications it is called a systematic cause and effect diagram; in this case we start with the desired effect (or objective) and then prepare the causes (or implementation steps).

The arrow diagram

The arrow diagram uses elements of the program evaluation and review technique (PERT) chart to organize daily operations in a sequence and to monitor its progress.

Process decision program chart (PDPC)

The PDPC is helpful for developing processes that do not fail. The tool studies all links in a given process, searching for weak ones. A good understanding of the failure modes at the weak links facilitates development of countermeasures to make the process strong and foolproof.

Matrix data analysis, or multivariate statistical analysis

The pictorial relationships of the matrix diagram can be quantified using appropriate multivariate statistical techniques. The multivariate techniques facilitate the understanding of complex relationships between variables resulting in well-informed business decisions.

Bibliography

Here is a list of recommended readings. Where appropriate, comments have been provided.

1. *Benchmarking: The Search for Best Practices That Lead to Superior Performance,* Robert C. Camp. ASQC Quality Press, 1989. A good book, the first we know of that provides a detailed discussion on the benchmarking process.
2. *Guide to Quality Control,* Kaoru Ishikawa, Asia Productivity Organization. Available from the ASQC Quality Press in the United States. Provides an excellent and easy to understand discussion on the seven quality control tools.
3. *Statistical Methods,* Hitoshi Kume, Association for Overseas Technical Scholarship. Provides an excellent and easy to understand discussion on the seven quality control tools.
4. *Statistical Quality Control Handbook,* Western Electric. An excellent, easy to read book on statistical quality control. The discussion focuses on control charts, design of experiments, and process capability.
5. *Statistical Quality Control,* E. L. Grant and R. S. Leavenworth, McGraw-Hill, New York. This is a good book that gives a detailed and theoretical discussion on statistical quality control. For a simpler and friendlier discussion, references 2, 3, and 4 are preferred. Many technical people, however, will prefer the technical discussion provided here.
6. *The Memory Jogger Plus,* M. Brassard, Goal/QPC. Provides a good discussion, with examples, on the new seven tools, also known as the seven new management tools.
7. *AT&T Reliability Manual,* D. J. Klinger, Y. Nakada, M. A. Menentez (eds.), Van Nostrand Reinhold, New York, 1990. Provides details on component derating and thermal design.
8. *Reliability Engineering for Electronic Design,* N. Fuqna, Marcel Dekker, New York, 1987. Provides details on component derating and thermal design.
9. *Handbook of Applied Thermal Design,* D. Brownell and E. Guyer, McGraw-Hill, New York, 1989. Provides details on component derating and thermal design.
10. *Practical Reliability Engineering,* P D. T. O'Conner, 2d ed., John Wiley, New York. Provides details on component derating, thermal design, and FMEA applications.
11. *How to Operate QC Circle Activities,* JUSE. This can be obtained from ASQC Quality Press in the United States.
12. *The Quality Circle Process: Elements for Success,* the ASQC Technical Committee, 1986. This can be obtained from the ASQC Quality Press in the United States.
13. *What Is Total Quality Control?,* Kaoru Ishikawa, Prentice Hall, Englewood Cliffs, N.J. Provides an excellent introduction, especially in concepts, but is less useful in implementing TQC.
14. *Total Quality Control,* A. V. Feigenbaum, McGraw-Hill, New York. Originally published in 1951. A very good book, its focus is on manufacturing and shop floor quality.
15. *Out of the Crisis,* Edward Deming, MIT Press. This book by the redoubtable Edward Deming explains his philosophy, via his 14 points. Not an easy read but nevertheless a must for all managers and professionals.

Index

Akao, Dr. Yoji, 136, 140
Akers, John, 94
American Express, 14, 24

Benchmarking, competitive (see Competitive benchmarking)
BMW, 43
Boeing Aircraft, 2, 112
Breakthrough objective, 50, 52, 58, 62, 64, 65, 79, 87, 90, 234, 235
Bridgestone Tire Company, 53
British Airways, 24
Business fundamentals plan (see Daily management plan)

CA-PDCA (Check, Analyze, Plan, Do, Check, Act) cycle, 97
CAR (Corrective Action Request), 21–22, 284
Cause and effect diagram, 66–67, 298–299
   for generating Hoshin plan strategies, 66–67
Chief executive's objectives, 64–65
Competitive benchmarking, 33–38, 39, 282, 283
   other thoughts on, 39
   process of, 34–38
   purpose of, 33–34
Component derating, 155–156
Control chart, 300–302
Control limits, 76–79, 88, 301–302
   purpose of, 78–79
Corrective action request (CAR), 21–22, 284
Cost:
   of losing a customer, 16
   managing of, 62, 64

Cost (Cont.):
   of poor quality, 95–96
Crosby, Phil, 6
Cross-functional teams/task force, 42, 69–72, 284
Customer, the high cost of losing one, 16
Customer complaints and feedback system, 17–27, 282, 283
   cue card for, 21
   feedback form (illus.), 20
   flowchart for, 19
   promoting and facilitating, 22–24
   trends in, 25–27
   what can you expect, 25
Customer needs, meeting and exceeding, 43–45, 46
Customer obsession, 13–15, 45–46
   pro-active activities, 15
   reactive activities, 15
   review of, 254–255
Customer satisfaction model, 30–32
   illustration, 31
   performance measures (table), 32
Customer satisfaction surveys, 27–30, 282
   setting priorities from, 29–30
Customers first, 282–283
Customers' voice, 140, 141, 142, 147–149
   developing the, 147–149

Daily management plan, 54, 56, 58, 74–79, 87, 88, 90, 257
   format and guidelines for, 74–78
Daily process management, 74, 76–77, 127–132, 202
   definition of, 128

308 Index

Daily process management (*Cont.*):
  in a factory, 129–130
  in marketing and sales, 130
  prerequisites for good, 130–132
  review of, 260–263
Deming, Edward, 1, 93, 96, 247, 306
Deming cycle, 96–97
Deming prize, 1, 242
  criteria for, 242–245, 246–247
Deming prize companies, profitability of, 3–4, 11*n*
Derating, of component, 155–156
Design standards, 154–160

Education:
  courses, 219–220
  of employees, 112–113, 219–220
  inadequate, 113
Employee education (*see* Education: of employees)
Employee participation, 205–206, 221–226
Employee suggestion scheme, 205, 211–219, 225, 226, 282
  flow chart of, 218
  form (illus.), 213–214
  grading system, 214–215
  guidelines, 211–216
  promotion of, 216–217
  reward system, 215
  what to expect, 219
Errors:
  human, 114
  problem solution model for, 114, 115
  the sources of, 114
Executive review, 274–278, 279
  agenda, 275

Fail-safing, 114
Failure Mode and Effects Analysis (*see* FMEA),
Feigenbaum, A. V., 8, 137, 205, 306
Fish-bone diagram, 299
Florida Power & Light, 100, 228
FMEA (Failure Mode and Effect Analysis), 124, 157–160, 184
  and the design process, 158–159
  benefits of, 157
  definition of, 157
  flow chart (illus.), 158
  form (illus.), 160

FMEA (Failure Mode and Eeffect Analysis) (*Cont.*):
  process of, 157–158
  requirements for, 159
Fool-proofing, 114

General Motors, 10, 284
Goal and control limits, 76, 78–79
Goals, setting of, 76, 78, 79–80, 112

Hewlett-Packard, 4, 39, 40, 91, 94, 100, 118, 187, 211, 215, 219, 229, 241, 252
Hewlett-Packard, Yokogawa, 4–6, 112, 114, 118, 228
  quality and productivity improvements, 4–6
Hokake, Takoshi, 282
Honda Motor Company, 51
Hoshin:
  plan, 53–91
    deployment matrix, 71–72
    format and guidelines, 58–72
    reviews, 56, 72, 80–85, 257, 284
    planning and daily management, relationship between, 58
    planning flowchart, 55
    planning process, 54, 66, 69
Hoshin Kanri, 50, 53, 57

IBM Corporation, 9, 94, 211, 215
Iizuka, Professor Yoshinori, 114, 116
Implementation plan, 54, 69, 71, 72–74, 80, 83, 84, 85, 89
  formats and guidelines, 72–74
Improvement:
  cycle, 93–95
    review of, 258–259
  plan, management of, 116
    illustration, 117
  of printed circuit production process, 118
  rate of, 112
  relationship between control and, 98–99
  targets for, 76–77, 112
Ishikawa, Kaoru, 8, 96, 111, 125, 305
Ishikawa diagram, 299
Issue list from previous year, 59, 62, 63
Itoh time management model, 233–236, 239

Juran, Joseph, 1, 6, 95, 137

Kano, Noriaki, 6, 7, 127–128
KJ (Kawakitu Jiro) method, 142, 144, 303
Komatsu flag system, 116
Kume, Hitoshi, 96, 114

Malcolm Baldrige award, 33, 242, 245–247
  criteria of, 245–246
Management by Objectives (MBO), 53, 57, 58
  and Hoshin Planning, 57, 58
Management kaizan (MK) teams, 207, 210, 224
Manager, basic activities, 233–236
Market place:
  success in, 39–40, 42
  failure in, 40–41
Marriot Corporation, 24
Matsushita Corporation, 51, 89, 205
MBO (Management by Objectives), 53, 57, 58
MBO and Hoshin Planning, 57, 58
Mercedes-Benz, 2–3, 43, 284
Miryokuteki hinshitsu, 282
MK (management kaizan) teams, 207, 210, 224
Morita, Akio, *xiii*
Motorola Company, 229, 230

NEC, 39, 89
Needs, customer, meeting and exceeding, 43–45, 46
New seven tools of quality control, 303–304
Nichijo Kanri, 53–54
Nordstrom, 44–45
Northern Telecom, 52–53
Numbers game, 79
Numerical targets, setting of, 79–80, 112

Objectives, 58–65, 68–71, 79, 80, 82–84, 87–91
  breakthrough (*see* Breakthrough objectives)
  chief executive's, 64–65
  deploying and cascading of, 68–72
  deploying in a large organization, 69–72
  types of, 62–64
Obsession, customer, 13–15, 45–46
Omark Industries, 154
Organizational learning, 283–284

Out of control report, 184–186
  illustration, 185

Pareto diagram, 297, 298
PDCA (Plan, Do, Check, Act) cycle, 66, 74, 80, 83, 96–111, 112, 114, 116–118, 120, 122, 124, 125, 136, 258, 259, 282, 283, 284
  benefits of, 99–100
  examples of, 100–111, 287–295
  in detail, 100–111
  uses of, 114, 116
  illustration, 98
Performance measures, 68
Peters, Tom, 44, 88, 89
PIMS (profit impact on marketing strategy) study, 3, 11*n*
Plan:
  annual, 50, 53, 87
  business fundamentals (*see* Daily management plan)
  daily management (*see* Daily management plan)
  implementation (*see* Implementation plan)
  long range, 50–53, 55, 89–90
  review of, 56, 72, 80–85
  (*See also* Hoshin plan)
Planning:
  calendar, 85
  issue list for, 59, 62–63
  process, 54, 66–69
    review of, 256–257
  putting it together, 85, 86, 87
Poka yoke, 114
Policy management, 50
Poor quality, the cost of, 95–96
Poor quality iceberg (illus.), 94
Postmortem:
  of project, 161–167, 178, 282, 283, 284
    format, 163–164
    meeting guidelines, 164–166
    objectives and benefits, 161–162
    process, 162
    sample of, 166–167
  for sales won and lost, 193–198, 282, 283, 284
    collecting information for, 196–197
    format, 195–196
    meeting guidelines, 164–166, 196
    objectives and benefits, 194
    process, 194–195

310 Index

Postmortem(*Cont.*):
    sample of, 197–198
Presidential audit, 274–278
    agenda, 275
Preventing problem recurrence, 120–121
Problem-solving hierarchy, 122, 124
Problems:
    preventing recurrence of, 120–121
    prevention by prediction, 122, 124
Process:
    concept, 128–129
    management, 128–133
        prerequisites for good, 130–132
    (*See also* Daily process management)
Processes:
    identifying key, 132–133
    in manufacturing environment,
        132–133, (illus.), 134
    prerequisites for good, 130–132
    in sales and marketing environment,
        132–133, (illus.), 135
    in software design environment, 199
    why manage, 129
Product:
    failure in marketplace, 40–41
    success in marketplace, 39–40, 42
    termination in design stage, 41–42
Product definition, 42, 133, 140
Product development process, 133
    (*See also* QFD)
Profit:
    of Deming prize companies, 3–4
    impact on marketing strategies, 3,
        11*n*
Profit and quality, 2–5
Project Sappho, 40
Publicity and promotion, 206, 208–210,
    215, 220–221
Purpose and vision, 51

QC (quality control) story, 97
QCDE (quality, cost, delivery, education), 62–64, 90
QFD (Quality Function Deployment),
    42–43, 46, 133, 136–154
    basics, 141–147, 203*n*
    benefits of, 137–139
    criteria for team members, 152
    definition of, 136
    facilitator for, 149–150
    increasing success of, 149–154
    introduction to, 136–139

QFD (Quality Function Deployment)
    (*Cont.*):
    nuts and bolts of, 139–141
    team models, 152
Quality:
    and profits, 2–5
    attractive, 6, 7, 43–45, 46
        in products and services, 43–45
    benefits of, 2–5
    circle evaluation system, 209–210
    circles, 206–211, 221–224, 225, 226,
        282
        education of, 207
        management of, 206–210
    costs, 95–96
    definition of, 6, 7
    education, 219–220
    first, 279–280, 282
    goals, 229–230
    house of, 141, 147
    must be, 6, 7
    revolution, 1–2, 9–11
    sacrificed, 186
    steering committee, 228–229
    teams, 206–211, 221–224, 225, 226, 282
        management of, 206–210
        (*See also* Quality circles)
    two dimensions of, 6, 7
    vision, 51
Quality assurance system, 178–186
Quality checkpoints chart, 181–184
Quality control (QC)
    new seven tools of, 303–304
    seven tools of, 11–112, 297–302
Quality, cost, delivery, education
    (QCDE), 62–64, 90
Quality Function Deployment,
    (*see* QFD)

Recognition, 206, 208–210, 215,
    220–221
Recurrence, prevention of, 120–121
Review, executive (*see* Executive review)
Root cause of problem generation, 100,
    105, 106, 108, 122, 125, 172
    table, 173

Sales:
    funnel, 187–188
    performance measures, 190–193
    process, 187, 189, 198
    situation, 190–191

Index

Sales (*Cont.*):
  won or lost postmortem, 193–198
  (*See also* Postmortem)
Sasaoka, Kenzo, 118
Satisfaction model, Customer (*see* Customer satisfaction model)
Seven new quality control tools, 303
  arrow diagram, 304
  KJ (Kawakita Jiro) method or affinity diagram, 303
  matrix data analysis, 303–304
  matrix diagram, 304
  process decision program chart (PDPC), 304
  relations diagram, 303
  systematic or tree diagram, 304
Seven quality control tools, 111–112, 297–302
  cause and effect diagram, 298–299
  checksheet, 297, 298
  control chart, 300–302
  graph and histogram, 299–300
  Pareto diagram, 297, 298
  scatter diagram, 300, 301
  stratification, 299
Shewhart, Walter, 96, 97
Shewhart cycle, 97
Siemens Company, 211, 225
Situation analysis, 68
Sony Corporation, *xiii*, 2, 39
Standards, 118, 120–122, 125, 178, 180, 181, 202, 260, 263
  benefits of, 120
  design, 158–160
  important points of, 120
  inadequate, 113
  sources for updates, 121–122
  update request (SUR), 121–122, 284
  illustration, 123
Status flag, 80–82
Strategies, 66–72
  generation of, 66–67
Suggestion scheme, 205, 211–219, 225, 226
  (*See also* Employee suggestion scheme)
Supplier management, 156–157
SUR (Standards update request), 121–122, 284
Surveys, Customer satisfaction (*see* Customer satisfaction surveys)

Targets, setting of, 79–80, 112

Teams:
  cross-functional, 42, 67–72, 284
  MK (management kaizan), 207, 210, 224
  quality, 206–211, 221–224, 225, 226, 282
Thermal design and measurement, 155
Time to market, 200–201
T-matrix (*see* T-type matrix)
Total participation, review of, 264–265
Total product concept (illus.), 44
Total Quality Control (*see* TQC)
Total quality creation, 232, 282
Toyota Motor Company, 22, 43, 51, 89, 284–285
TQC (Total Quality Control):
  and business success, 2–5
  and efficiency, 284
  and profits, 2–5
  and time management, 233–237
  audit, 241
  benefits of, 5, 10–11
  creativity and business success, 285
  definition of, 5–6
  diagnosis, 241
  elements of, 8–9
  essence of, 281–285
  generic model of, 2–3
  genesis of, 2–3
  getting started in, 227–233
  headquarters, 230
  objective of, 5–6
  phases of, 230–233
  review, 241, 247–279
    agenda, 248–250
    checklist, 248, 252–266
    objective, 247
    process, 247–248
    recommendations and report, 269
    scoring system, 250–266
    team, 267
    techniques, 267–268
    weaknesses, 269–274
Training (*see* Education)
T-type matrix, 161, 167–178, 303–304
  analysis and interpretation of, 172–176
  application of, 174
  mechanics and preparation of, 168–172
  objectives of, 168
  when and how to use the, 176–177

Union Carbide, 41–42

Vision, purpose and, 51

Wall Street Journal, 13
Whispering of Satan, 282

Xerox Company, 9, 33, 229–230

Yokogawa Hewlett-Packard, 4–6, 112, 114, 118, 228
 quality and productivity improvements, 4–5
Young, John, 4, 13

Zero defects, 114, 201–202

ABOUT THE AUTHOR

Sarv Singh Soin is currently operations manager for Hewlett-Packard's Asia Pacific Distribution Operation, based in Singapore. Prior to this he was Quality Manager for Hewlett-Packard's Intercontinental Operations, based in Palo Alto, California. Altogether he has been a quality manager for about 9 years in manufacturing as well as in sales at Hewlett-Packard.